Telecommunicatio
Principles

D0727636

JOIN US ON THE INTERNET VIA WWW, GOPHER, FTP OR EMAIL:
WWW: http://www.thomson.com
GOPHER: gopher.thomson.com
FTP: ftp.thomson.com
EMAIL: findit@kiosk.thomson.com

A service of I(T)P

TUTORIAL GUIDES IN ELECTRONIC ENGINEERING

Series editors
Professor G.G. Bloodworth, *University of York*
Professor A.P. Dorey, *University of Lancaster*
Professor J.K. Fidler, *University of York*

This series is aimed at first- and second-year undergraduate courses. Each text is complete in itself, although linked with others in the series. Where possible, the trend towards a 'systems' approach is acknowledged, but classical fundamental areas of study have not been excluded. Worked examples feature prominently and indicate, where appropriate, a number of approaches to the same problem.

A format providing marginal notes has been adopted to allow the authors to include ideas and material to support the main text. These notes include references to standard mainstream texts and commentary on the applicability of solution methods, aimed particularly at covering points normally found difficult. Graded problems are provided at the end of each chapter, with answers at the end of the book.

Transistor Circuit Techniques: discrete and integrated (2nd edn) — G.J. Ritchie
Computers and Microprocessors: components and systems (2nd edn) — A.C. Downton
Telecommunication Principles (2nd edn) — J.J. O'Reilly
Digital Logic Techniques: principles and practice (2nd edn) — T.J. Stonham
Signals and Systems: models and behaviour — M.L. Meade and C.R. Dillon
Electromagnetism for Electronic Engineers — R.G. Carter
Power Electronics — D.A. Bradley
Semiconductor Devices: how they work — J.J. Sparkes
Electronic Components and Technology: engineering applications — S.J. Sangwine
Control Engineering — C. Bissell
Software Engineering — D.C. Ince
Electronic Product Design — T. Ward and J. Angus

Telecommunication Principles

Second edition

J.J. O'Reilly
School of Electronic Engineering Science
University of Wales, Bangor

CHAPMAN & HALL
University and Professional Division
London · Weinheim · New York · Tokyo · Melbourne · Madras

Published by Chapman & Hall, 2–6 Boundary Row, London SE1 8HN, UK

Chapman & Hall, 2–6 Boundary Row, London SE1 8HN, UK

Chapman & Hall GmbH, Pappelallee 3, 69469 Weinheim, Germany

Chapman & Hall Inc., One Penn Plaza, 41st Floor, New York, NY10119, USA

Chapman & Hall Japan, Thomson Publishing Japan, Hirakawacho Nemoto Building, 6F, 1-7-11 Hirakawa-cho, Chiyoda-ku, Tokyo 102, Japan

Chapman & Hall Australia, Thomas Nelson Australia, 102 Dodds Street, South Melbourne, Victoria 3205, Australia

Chapman & Hall India, R. Seshadri, 32 Second Main Road, CIT, East, Madras 600 035, India

First edition 1984
Reprinted 1985, 1987, 1989
Second edition 1989
Reprinted 1991, 1992, 1993, 1994, 1996, 1998

© 1984, 1989, J.J. O'Reilly

Printed in China

ISBN 0 412 43700 7

Contents

Preface   vii

1 Signals, systems and communications   1
  Communication signals   2
  Communication channels   4
  Communication networks   14
  Telecommunications worldwide   20
  Summary   21
  Problems   22

2 Signal representation and analysis   23
  The time domain   24
  The frequency domain   28
  Fourier series analysis   34
  Frequency domain representation of aperiodic signals   38
  Fourier transforms   39
  Frequency domain representation for signals of arbitrary waveshape   48
  Amplitude distribution of signals   50
  Noise processes   53
  Summary   55
  Problems   55

3 Sinusoidal carrier modulation   58
  Introduction   58
  Amplitude modulation   58
  Angle modulation   66
  Communication in the presence of noise   69
  Frequency division multiplexing   74
  Summary   75
  Problems   76

4 Radio receiver principles   78
  Tuned radio frequency (TRF) receiver   80
  Superheterodyne (superhet) receivers   81
  Summary   86
  Problems   86

5 Pulse modulation systems   87
  Pulse amplitude modulation   87
  Other pulse modulation schemes   93
  Time division multiplexing   94
  Summary   95
  Problems   96

6 Pulse code modulation 97

Quantization 97

Sampling and pulse encoding 99

Non-uniform quantization 101

Differential pulse code modulation 103

PCM–TDM telephony 105

Summary 106

Problems 107

7 Digital communications 108

Digital transmission 108

The eye diagram 113

Signal design 113

Error probability 116

Coding for digital transmission 119

Digital modulation 127

Summary 128

Problems 129

8 Systems case studies 131

Broadcast FM radio 131

Television systems 134

Videotex systems 141

The compact optical disc as a communication system 146

Summary 148

Appendix A: Decibels 149

Appendix B: Some Fourier transform results 151

Answers to numerical problems 152

Index 155

Preface

This book provides a first introduction to the subject of telecommunications suitable for first and second year undergraduates following degree or similar courses in electronic engineering. There are very few specific prerequisites other than a general background in electric circuit principles and a level of mathematical maturity consistent with entry to engineering courses in British universities.

The intention is to provide a broad perspective of modern telecommunication principles and applications. Following a general overview of telecommunications, a thorough, albeit introductory, treatment is provided of underlying principles such as signal representation and analysis, sampling, analogue and digital transmission, modulation and coding. The book concludes with a description of several important systems applications which serve as case studies to illustrate further the principles introduced and demonstrate their application in a practical context.

Many people have contributed, directly and indirectly, to this book. I am especially grateful to Professor Kel Fidler of the Open University for suggesting that I write the book and for the support and guidance he has provided throughout the endeavour. The Telecommunications Research Group of the Department of Electrical Engineering Science at the University of Essex provided a stimulating environment in which to develop my appreciation of telecommunication systems and in particular Professor Ken Cattermole influenced greatly my thinking.

A close working relationship with British Telecom Research Laboratories has proved most fruitful and I am indebted to Drs Peter Cochrane, Ian Garrett and Mike O'Mahony of BTRL for valuable discussions. Dr Tim Dennis and his colleagues in the Visual Systems Research Group at Essex provided invaluable assistance with television picture coding for the illustrations, and special thanks go to Jenny, my photographic model.

In preparing this new edition I have sought to preserve the introductory nature of the text whilst at the same time strengthening the formal mathematical foundations provided in Chapter 2. In particular, the treatment of the Fourier transform has been extended and greater emphasis has been placed on the influence of noise on system performance.

Signals, Systems and Communications   1

Objectives

☐ To provide a broad overview of telecommunication systems.

☐ To introduce the concept of a signal and show the relation to electrical waveforms.

☐ To introduce the concept of a communication channel and to discuss diverse practical realizations, such as; coaxial cables, optical fibres and radio links.

☐ To introduce communication networks and discuss different topologies and design philosophies.

☐ To explain the terms local area network (LAN), wide area network (WAN) and integrated services digital network (ISDN)

Telecommunication systems are concerned with the transmission of information, or *messages*, from one point to another. The origin of the message is referred to as the *source* and the destination as the *sink*. An illustrative communication link is shown in Fig. 1.1. In order to effect this transfer of information the messages from the source are represented by a *signal* which takes the form of a voltage varying with time: the signal waveform. Arguably this is a restricted view of telecommunications, which means simply *communication at a distance*, and could reasonably be taken to encompass other forms of message transport, of which the postal service and acoustic transmission systems, such as the speaking tube of the nineteenth century, are but two examples. The restriction is readily justified, though, since electrical — or more properly electromagnetic — signals are invariably used for near instantaneous long-range communications. The main reasons for this are that electrical signals are readily generated and modified with the aid of electronic circuits and they travel at, or close to, the velocity of light.

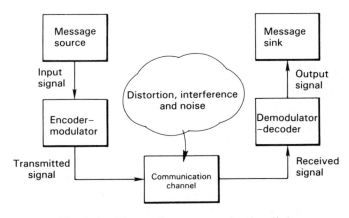

Fig. 1.1   Illustrative communication link.

Exercise 1.1 Determine the time delay experienced by:
(a) an electrical signal travelling at the velocity of light in vacuum (approximately 3×10^8 m/s)
(b) an acoustic signal travelling at the velocity of sound in air (approximately 343 m/s)
over a distance of 5000 km, corresponding approximately to transatlantic communication.

Considering only signal delay, comment briefly on the viability of some form of acoustic transmission for two-way long distance telecommunication.

Communication Signals

An example of an electrical signal is the voltage waveform produced by a microphone in response to a spoken message. The precise form of the speech waveform depends on the message, the speaker and the characteristics of the microphone but an illustrative example is shown in Fig. 1.2a. The voltage varies

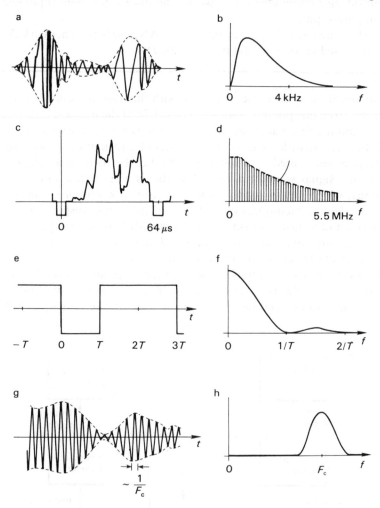

Fig. 1.2 Typical signal waveforms and corresponding power spectra. (a) Speech waveform; (b) speech spectrum; (c) television waveform; (d) television spectrum; (e) binary data waveform; (f) binary data spectrum; (g) bandpass signal; (h) bandpass spectrum.

continuously with time in an unpredictable manner and it is the task of the communication link to produce a close replica of this waveform at the receiver. This may be achieved by passing the signal along a pair of wires but communication over long distances usually requires some modification of the signal to render it compatible with an available *channel* such as a radio link. Depending on its form this modification is referred to variously as encoding and/or modulation and a reverse process is required at the receiver to *undo* the modification and recover the original message waveform. These processes are discussed in some detail in later chapters; for the present consider the message signal itself. In practice it is neither necessary nor possible for the message waveform to be reproduced *exactly* at the receiver; a close approximation suffices. This last observation is of profound importance and influences significantly the design of communication systems. For example, a speech waveform may be passed through a filter which removes certain frequency components and yet the speech may be quite intelligible. Some idea can be obtained of which components are likely to be most important by examining the distribution of the signal power with frequency, as shown in Fig. 1.2b. This form of signal representation, known as a *power spectrum* or *power spectral density function*, is discussed further in Chapter 2. Here it is sufficient to note that a communication system should provide good transmission at frequencies for which the signal power spectrum is significant. Just what level constitutes *significant* in this context is open to question; any judgement as to whether or not a speech signal has been significantly impaired by filtering must be *subjective*. The quality of a communication channel may need to be assessed on the basis of subjective tests. From such tests it has been found that if a communication channel passes relatively unimpaired signal components with frequencies in the critical range 300 Hz to 3.4 kHz then speech is entirely intelligible. In view of this, telephone systems incorporate filters to limit the speech signal to this frequency band. To allow for imperfections of practical filters telephone speech is often treated as if it were 4 kHz lowpass. That is, the constituent signal frequencies are taken to extend from very low values (essentially d.c.) up to 4 kHz. We refer to the range of significant frequency components in a signal waveform as the signal bandwidth.

Speech, albeit very important, is only one type of signal it is wished to convey. High quality music signals contain significant frequency components over the whole of the audible range which extends from about 20 Hz to approaching 20 kHz. Good reproduction is obtained by treating music signals as lowpass with a bandwidth of 15 kHz.

Television provides another example. The saying 'a picture is worth a thousand words' is exemplified by the bandwidth requirements of 5.5 MHz lowpass for broadcast quality television signals. Fig. 1.2c provides a representative illustration of a television waveform and a television system forms the basis of one of the case studies of Chapter 8.

A binary data signal is shown in Fig. 1.2e. The voltage varies with time, alternating between two values to delimit the data corresponding with the message. The smallest interval between transitions, T, is known as the signalling interval or bit time and $1/T$ is the signalling rate in bits per second[bit/s]. It can be seen from the power spectrum that most of the signal power is concentrated at frequencies less than $1/T$. Therefore, it can be expected that the transmission bandwidth required for a binary data system would be proportional to, and perhaps of the same order as, the bit rate. More is said about binary data signals in Chapter 7

The power spectrum of a signal is discussed further in Chapter 2.

For an extensive treatment of speech signal characterization and processing in the context of telephony see Richards, D.L., *Telecommunication by Speech*, Butterworth, 1973.

where these tentative conclusions are found to be broadly correct. Here it is noted that if there is a requirement to send a large amount of data in a short time a high data rate is involved and thus a wide bandwidth channel is required. It is quite common to encounter data rates up to 140 Mbit/s in the telecommunication network, with experimental systems operating at rates of several Gbit/s being studied at the time of writing (1988).

The signals discussed so far are all lowpass. That is, they have been concentrated in a frequency band extending from close to d.c. up to some maximum frequency f_{max}, which thus corresponds closely to the signal bandwidth, $B \simeq f_{max}$. Not all signals are of this type. Some are bandpass, concentrated in a band of frequencies extending from a lower frequency f_{min} to an upper frequency f_{max} and have a bandwidth $B \simeq f_{max} - f_{min}$. A bandpass signal is illustrated in Fig. 1.2g. Most commonly, bandpass signals are derived from lowpass signals by way of a modulation process. The purpose of this lowpass to bandpass signal transformation is to enable the message to be conveyed over a bandpass channel, a topic examined in Chapter 3.

Having discussed briefly some of the types of signal it may be required to transmit and noted the wide diversity in terms of signal bandwidth, attention is now turned to the possible channels over which these signals may be conveyed.

Communication Channels

An ideal communication channel would convey unimpaired the communication signal from source to destination. To do this it would have to pass all frequencies equally well; that is, it would have a uniform frequency response. Also, it would need to be free from extraneous disturbances such as unwanted signals which might be added to the wanted signal and interfere with communication. And, of course, it would be ideal for the signal to reach its destination unattenuated to obviate the need for signal amplification at the receiver. It comes as no suprise that practical channels are not like this! A signal may suffer considerable attenuation, which may be more severe at some frequencies than others and may also vary with time. In addition, the signal is liable to suffer corruption by the addition of interfering signals. The relative significance of these various forms of signal impairment depends on the particular physical channel involved. Consider briefly two broad classifications: (i) guided wave systems in which the signal is conveyed via some constraining physical medium such as a pair of wires, and (ii) radio systems in which signal transfer is effected via a freely propagating electromagnetic wave.

Electromagnetic theory is discussed in two further texts in this series: Compton, A.J., *Basic Electromagnetism and its Applications* and Carter, R.G., *Electromagnetism for Electronic Engineers*, both Van Nostrand Reinhold, 1986.

First note that the range of frequencies available is very wide, ranging from d.c. (zero frequency) to optical frequencies of the order 10^{14} Hz as shown in Table 1.1. In order to exploit effectively this wide range of frequencies it is necessary that source messages be processed and combined with other messages to form a single composite signal in which the constituents are individually identifiable. This process, known as *multiplexing*, is discussed in later chapters. For the present it suffices to note that message signals can be combined and translated in frequency for transmission and that the individual messages are recoverable at a remote location.

Table 1.1   Communication Frequency Designations

Baseband:	a basic message signal concentrated at low frequencies with bandwidth depending on type of message		
		Frequency	*Wavelength*
VLF:	very low frequencies	<30 kHz	10 km
LF:	low frequencies	300 kHz	1 km
MF:	medium frequencies	3 MHz	100 m
HF:	high frequencies	30 MHz	10 m
VHF:	very high frequencies	300 MHz	1 m
UHF:	ultra high frequencies	3 GHz	10 cm
SHF:	super high frequencies	30 GHz	1 cm
EHF:	Extra high frequencies	300 GHz	1 mm
(millimetre waves)			
lightwaves	Optical frequencies	$\gtrsim 10^{14}$ Hz	1 μm

Twisted Pair Cables

A very widely used form of communication channel consists simply of a pair of electrical conductors. Such *wire pairs* may be used, for example, to connect computer terminals to a nearby central processor or to connect telephone instruments to a local exchange. The physical structure of the *transmission line* strongly influences the *primary electrical parameters*:

More details of transmission line principles may be found in Coates, R.F., *Modern Communication Systems*, Macmillan, 1975, 1983.

capacitance between conductors	C [F/m]
conductor resistance	R [Ω/m]
leakage conductance between conductors	G [S/m]
inductance	L [H/m]

These in turn influence the characteristics of the transmission line as a communication channel. An analytic treatment of line operation is not considered here, but it should be noted that the total attenuation of a cable depends on its length. Consider, for example, two identical 1 metre sections of cable connected in cascade, as shown in Fig. 1.3, with the second section appropriately terminated. The voltage transfer ratio for each section is given by

Here it is assumed that the second section is appropriately terminated so that 'looking into' section 2 or section 1 towards the load the same impedance, Z, is *seen*.

$$V_1/V_0 = K; \qquad V_2/V_1 = K \tag{1.1}$$

Hence the overall voltage transfer ratio is

$$V_2/V_0 = V_2/V_1 \cdot V_1/V_0 = K^2 \tag{1.2}$$

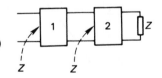

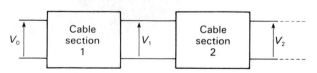

Fig. 1.3   Cascade connection of cable sections.

More generally, for a cable of length d the transfer ratio is

$$V_d/V_0 = K^d \qquad (1.3)$$

Decibel measures are reviewed briefly in Appendix A.

Expressing this in terms of decibels:

$$\begin{aligned}
20 \log_{10}(V_d/V_0) &= 20 \log_{10}(K^d) \\
&= d20 \log_{10}(K) \\
&= -\alpha d \ \mathrm{dB}
\end{aligned} \qquad (1.4)$$

Here $\alpha = -20 \log_{10}(K)$ is the cable attenuation in dB/m and the introduction of the minus sign yields a positive value for the cable attenuation constant given that $K < 1 \Rightarrow \log(K) < 0$. The signal attenuation in decibels is thus proportional to transmission distance d for a cable system.

As one might expect, given the primary electrical parameters noted above, the attenuation is also frequency dependent and tends to increase with frequency as shown in Fig. 1.4a. Hence, to achieve an acceptable bandwidth some form of frequency sensitive compensating network may be required. Such a network is known as an equalizer since it makes uniform, or equal, the transmission of the cable over the frequency range of interest. For simple pair cable this can often be achieved by introducing series inductors, known in this context as *loading coils*, at suitable points along the cable. The effect of loading coils on the transmission characteristic of a length of telephone cable is also illustrated in Fig. 1.4a. Notice that the resulting bandwidth is rather limited and depends on length.

A representative multi-pair telephone cable is shown in Fig. 1.4b from which a further deficiency may be anticipated. Currents flowing in the conductors produce magnetic fields which can couple to, and induce currents in, other wire pairs in the same cable. Cross-coupling may also arise due to capacitance between the different conductors. The result is that a wanted signal may be impaired by the addition of interfering signals from other cable pairs. The problem was first encountered with speech telephony where it takes the form of talking in the background — it is thus referred to as *crosstalk*. The various wire-pairs may be randomized in position along the length of the cable thus ensuring that a given channel receives interference more or less uniformly from all other channels rather than predominantly

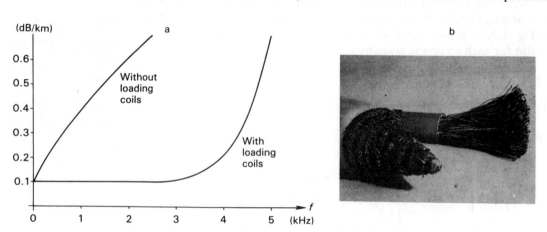

Fig. 1.4   (a) Attenuation characteristic of twisted-pair cable; (b) multi-pair cable.

from one. This has the effect of rendering the crosstalk less intelligible. Subjective tests have shown that telephone users are considerably more tolerant to crosstalk if it is unintelligible.

Coaxial Cables

For applications requiring greater bandwidths than are available with wire pairs coaxial cables are generally used. The attenuation for such a cable increases approximately with the square root of frequency so equalization is required for long distance, wide bandwidth operation. Coaxial cable systems are available which provide a usable signal bandwidth of 60 MHz or can accommodate 140 Mbit/s digital transmission. Greater bandwidths/data rates can be achieved but large cable diameters are required for low transmission loss to result. This arises because at high frequencies the current is concentrated near the surface of a conductor. Owing to this so called *skin effect* an increase in cable diameter is required as the operating frequency is increased if the effective cross-sectional area of the conductor is to be maintained. A further advantage of the coaxial structure is that it can result in much reduced crosstalk. The electric and magnetic fields are almost entirely constrained within the cable, the outer, surrounding conductor being earthed.

Optical Fibres

Guided wave transmission at microwave frequencies is not currently employed for long distance communications despite an extensive research and development effort spread over several years, in the UK, USA and elsewhere. This is largely because recent advances in optical (or lightwave) communications have made this latter option far more attractive both technologically and economically. Guided wave optical communication systems make use of fine strands of high purity glass, approximately 100 microns (μm) in overall diameter. The technology is particularly well suited to digital communications so we will concentrate on this. Note in passing, however, that analogue communication over optical channels is possible both directly and with the aid of various pulse modulation schemes (see Chapters 5, 6, and 7).

The use of optical fibres for communications was proposed in 1966 by two research engineers, Kao and Hockham, working at the British laboratories of STL.

The simplest form of optical fibre has a relatively large central core ($\sim$ 50 μm in diameter) with refractive index n_1 surrounded by a lower refractive index cladding n_2. Light can thus travel along the central core by way of a series of total internal reflections at the core–cladding interface, as shown in Fig. 1.5. In order to be

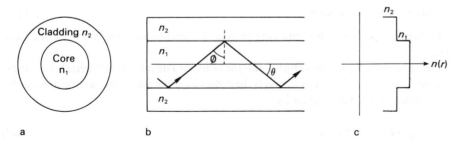

Fig. 1.5 Step-index multimode optical fibre. (a) End view; (b) side view showing ray propagation; (c) refractive index profile.

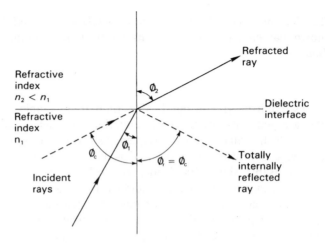

Fig. 1.6   Total internal reflection.

trapped within the core in this way light rays must be at a relatively shallow angle relative to the fibre axis. This may be appreciated as follows:

Consider light undergoing refraction at a plane dielectric interface, as shown in Fig. 1.6. Snell's law for refraction takes the form

$$n_1 \sin \phi_1 = n_2 \sin \phi_2 \tag{1.5}$$

Here $n_1 > n_2$ implies $\phi_1 < \phi_2$. If the angle of incidence ϕ_1 is increased a critical angle is reached, defined by

$$n_1 \sin \phi_c = n_2 \tag{1.6}$$

That is, Snell's law predicts that the refracted ray makes an angle of 90° to the perpendicular. The ray is totally reflected back into the high index region; total internal reflection occurs and this applies also for all angles $\phi_1 > \phi_c$. The angle θ of the ray to the fibre axis is related to the angle at the core–cladding interface by

$$\sin \phi = \cos \theta \tag{1.7}$$

and ϕ_c defines the minimum value for ϕ_1 for total internal reflection. This corresponds to a maximum value θ_m for θ:

$$\sin \phi_c = n_2/n_1 = \cos \theta_m \tag{1.8}$$

Light is trapped within the fibre core provided it makes an angle $\theta < \theta_m$ to the fibre axis.

Consider now a very short pulse of light launched into a fibre of length L at angle θ to the axis. If $\theta = 0$ the light travels directly along the fibre core axis and emerges at the exit after a delay given by

$$\tau_{min} = L/v_1 = n_1 L/c \tag{1.9}$$

where $v_1 = c/n_1$ is the velocity of light in the core. On the other hand, light inclined at an angle θ to the axis follows a zig-zag path of length P given by

$$P = L/\cos \theta \qquad (1.10)$$

and so arrives at the exit after a delay which depends on θ:

$$\tau(\theta) = P/v_1 = \frac{n_1 L}{c \cos \theta} \qquad (1.11)$$

For the extreme ray, travelling at angle θ_m,

$$\tau_{max} = \frac{n_1 L}{c \cos \theta_m} = \frac{n_1^2 L}{c n_2} \qquad (1.12)$$

Hence if a short pulse of light is launched such that components enter the fibre at all angles $0 < \theta < \theta_m$ the pulse arrives at the exit spread out in time by an amount given by

$$\Delta\tau = \tau_{max} - \tau_{min} = \frac{n_1 L}{c}\left(\frac{n_1}{n_2} - 1\right) = \frac{n_1 L}{n_2 c}(n_1 - n_2) \qquad (1.13)$$

This *pulse spreading* places a limit on the rate at which pulses may be transmitted over a fibre of given length, L. Pulse spreading of this sort is characteristic of step index fibres, that is, fibres in which the refractive index is uniform in the core and steps down abruptly to the cladding value.

An optical fibre has core refractive index $n_1 = 1.5$ and cladding refractive index $n_2 = 1.45$. Estimate the maximum signalling rate attainable for fibre lengths of (i) 100 m (ii) 10 km.

Worked Example 1.1

Solution: To avoid undue overlap of pulses at the exit of the fibre they must be separated in time by approximately $\Delta\tau$ or more. Hence the maximum signalling rate is $f_{max} \simeq 1/\Delta\tau$. Now $n_1 \simeq n_2$; hence

$$\Delta\tau = \frac{n_1}{n_2}\frac{L}{c}(n_1 - n_2) \simeq \frac{L(n_1 - n_2)}{c} = 0.05L/c$$

whence

$$f_{max} \simeq c/0.05L \qquad (1.14)$$

(i) For $L = 100$ m

$$f_{max} \simeq 3 \times 10^8/(0.05 \times 100) \simeq 60 \text{ MHz}$$

(ii) For $1 = 10$ km

$$f_{max} \simeq 3 \times 10^8/(0.05 \times 10^4) \simeq 600 \text{ kHz}$$

Notice that the effective bandwidth of a step index fibre depends on length L and on the index difference $(n_1 - n_2)$. For short haul systems, for example data links within buildings, ships, aircraft and so on, fibres of this type are quite suitable. For long haul, high data rate systems, however, alternative structures are required.

If the refractive index in the core region gradually reduces from a maximum at the centre to some lower value at the cladding, as shown in Fig. 1.7, the fibre bandwidth is greatly improved. The light now follows a curved, approximately sinusoidal path, being continuously refracted within the core rather than reflected at the core–cladding boundary. A fibre of this type is referred to as a *graded index fibre*. Notice that light rays at shallow angles, with the shorter physical path lengths, are confined to the central high index region where the velocity of light is relatively low. In contrast, rays at steep angles with long physical path lengths penetrate further into the low refractive index regions where the velocity of light is higher. Appropriate choice of the refractive index profile results in all ray paths having almost the same propagation delay and pulse spreading is very much reduced compared with a step index fibre. The best index profile has been found to be approximately parabolic, although there is no profile which results in all ray paths having *precisely* the same delay. Hence some pulse spreading is inevitable and can prove a limiting factor for high data rate, long distance systems.

In these discussions of optical fibres *rays* of light have been considered; we have used a *ray optics* model. This is reasonable provided the fibre diameter is large compared to the wavelength of light so that diffraction/wave effects can be ignored. More properly, though, it should be noted that light is an electromagnetic phenomenon. If wave effects are included in the analysis it emerges that light cannot propagate in a fibre at all the angles predicted by ray optics but only at certain discrete angles. These ray angles correspond to different modes of electromagnetic propagation. For large core diameter fibres with a large index difference there are many possible propagation angles or modes and such fibres are referred to as *multimode*. The pulse spreading corresponds to the different modes having different propagation velocities. In a graded index multimode fibre the mode velocities are almost, but not quite, equal. With a large core diameter and index difference the allowed ray angles are very closely spaced and the continuous approximation of ray optics is reasonable. However, by making the core diameter and index difference sufficiently small we can ensure that there is essentially only one possible mode. The result is a *monomode* fibre and since there is only a single mode pulse spreading due to mode velocity differences cannot exist.

There are other effects which can limit the bandwidth of an optical fibre system. For example, the refractive index varies with wavelength so if the optical source emits more than one wavelength pulse dispersion may result owing to different wavelengths travelling at different velocities. This has resulted in very long haul

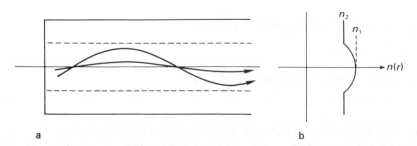

Fig. 1.7   Graded index multimode fibre. (a) Side view showing ray propagation; (b) refractive index profile.

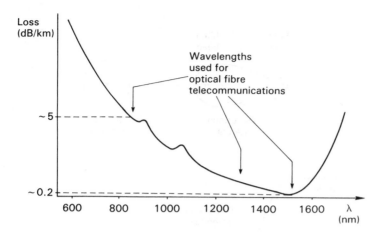

Fig. 1.8 Illustrative loss versus wavelength characteristic for silica optical fibre.

optical fibre systems being designed to operate at a wavelength of 1300 nm where material dispersion is minimal for silica — the dominant component of high grade optical fibres. The attenuation at this wavelength is also very low, but it can be even lower at 1550 nm, as shown in Fig. 1.8. Fortunately it is possible to achieve both low loss and low dispersion at 1550 nm by appropriate monomode fibre design. At the time of writing, optical fibre systems operating at 850 nm, 1300 nm and 1550 nm are in use and are being further developed. A particularly intriguing aspect of optical fibre communication is that this one basic technology is applicable to such a wide range of applications, from low data rate, short distance links within equipments and buildings to very high bandwidth, long distance links exemplified by transatlatic submarine telecommunications systems.

Certain other aspects of optical fibre systems and technology, as well as optoelectronics more generally, are discussed in another text in this series: Watson, J., *Optoelectronics*, Van Nostrand Reinhold, 1986.

Radio Systems

Considering radio systems, an electromagnetic wave is launched from an antenna at the transmitter, propagates through the atmosphere and on arrival at the receiver is picked up by a receiving antenna. Depending on the form of the antenna, the wave may propagate out from the transmitter in all directions or may be restricted to some narrow sector in a chosen well-defined direction. In the first case the transmitting antenna is said to be *omnidirectional*, as opposed to *directional*. Omnidirectional transmission is used for radio broadcasting since receivers may generally be positioned at any location around the transmitter. On the other hand, if it is wished to implement a point-to-point radio communication link then use of a *directional* antenna is desirable since this causes the available power of the transmitter to be concentrated in the direction of the receiver. For either system it is appropriate for the receiver antenna to be directional since then the receiver is most sensitive to the desired signals as opposed to potential interference signals arriving at the receiver location from other directions. Illustrative antenna directionality patterns are shown in Fig. 1.9. The directionality of an antenna is related to the ratio of the antenna physical dimension, D, to the wavelength λ. A directional antenna is often referred to as having a gain G since the concentration of the transmitted power in a given direction results in an increased signal power density compared with an omnidirectional antenna. For example, for a parabolic reflector

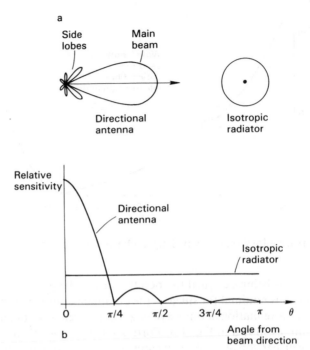

Side lobes

Main beam

Directional antenna

Isotropic radiator

Relative sensitivity

Directional antenna

Isotropic radiator

0 $\pi/4$ $\pi/2$ $3\pi/4$ π θ

b

Angle from beam direction

Fig. 1.9   Antenna directionality. (a) Polar diagrams; (b) relative sensitivity versus angle.

antenna such as might be used at microwave frequencies the power gain G is given by

$$G \simeq (D/\lambda)^2 \tag{1.15}$$

where D is the diameter of the reflecting parabolic dish.

Worked Example 1.2 Consider a 2 m diameter reflecting parabolic dish used as the transmitting antenna for (a) $\lambda = 10$ cm and (b) $\lambda = 3$ cm microwave radio systems. (i) Determine the antenna gain in each case and (ii) comment on the increased effective radiated power provided by (b) compared with (a).

Solution:
(i)   (a) $G \simeq (2/0.1)^2 = 400 \equiv 26$ dB
      (b) $G \simeq (2/0.03)^2 = 4,444 \equiv 37$ dB

(ii) There is thus an effective increase in radiated power of approximately 37 − 26 dB = 11 dB owing to the reduced wavelength and correspondingly improved directionality.

Any practical antenna cannot be 100% efficient. In the case of a directional antenna this may be partly owing to imperfect radiation efficiency and partly to geometrical irregularities giving rise to some radiation outside the intended beam directions. These factors are often accounted for by referring to the *effective isotropic radiated power* (e.i.r.p.). The radiation from a directional antenna is thus equivalent to $G \times$ (e.i.r.p.) from an omnidirectional antenna.

No matter how directional the antenna, as the wave moves away from the transmitter it spreads out. For the omnidirectional case a good analogy, albeit in two dimensions rather than three, is provided by the waves caused by a stone thrown into a pool. By the e.i.r.p. concept a similar argument holds for the directional case too. As a consequence, the power density reduces with increasing distance from the transmitter. To quantify this, consider concentric spheres of radius r_1, r_2 centred on the transmitter. The surface area of a sphere varies with $1/r^2$ but the power flow across the total area of each sphere, assuming no power dissipation, is the same. The power density thus reduces proportional to $1/r^2$ and the power detected by a given receiver at distance r_2 compared with power detected by the same receiver at distance r_1 is given by

$$Pr_2/Pr_1 = (r_1/r_2)^2 \equiv 20 \log_{10}(r_2/r_1) \text{ dB} \tag{1.16}$$

It is concluded the effective signal attenuation in decibels for such a radio system varies with the log of the distance between the transmitter and the receiver. It is interesting to contrast this with a cable system for which, as has been seen, the attenuation in decibels increases linearly with distance. On these grounds it is tempting to suggest that radio systems are best suited to long distance communications since, as shown in Fig. 1.10, the logarithmic dependence of loss means that ultimately, at some distance, the loss of a radio system is less than the loss of a cable system. While there is some truth in this, the argument is too trite. To be practically useful any such comparison must take account of different forms of radio wave propagation and of actual cable attenuation levels.

The form of propagation discussed above is appropriate to 'line of sight' systems in which the transmitter and receiver antennas are in view of one another. There are, however, other forms of radio propagation which find application. Briefly, the most important forms are as follows:

(i) *Surface wave or ground wave.* The radio wave travels along the surface, following the curvature of the Earth, as a result of currents flowing in the ground. At low frequencies this is the dominant propagation mechanism and it can provide

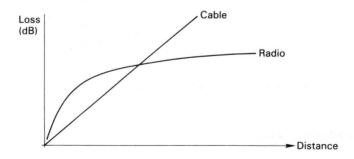

Fig. 1.10   Comparison of linear (cable) and logarithmic (radio) loss characteristics.

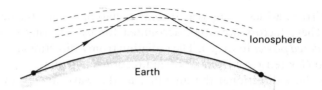

Fig. 1.11   Ionospheric propagation.

for long distance communications. It is, however, prone to variation with the conductivity of the ground.

(ii) *Ionospheric propagation.* Radiowaves can be refracted by the ionosphere — a layer of ionized particles above the surface of the Earth — and returned to the surface some considerable distance from the transmitter. The physical mechanism whereby the radiowaves are bent and returned to Earth is one of continuous refraction due to gradual reduction with height of the refractive index in the ionosphere, as shown in Fig. 1.11. The height and density of the ionosphere depends on solar activity and there is considerable variation between day and night and with the seasons. The wave returned to earth may be reflected at the surface so that very long distance communication via multiple hops is possible.

(iii) *Tropospheric scattering.* Radio waves can be scattered by small particles in the lower atmosphere to provide over-the-horizon radio communication as shown in Fig. 1.12.

(iv) *Free space propagation.* A freely propagating radio wave provides for line of sight communications. Such a wave can propagate without the aid of a physical medium and this mode is thus known as free space propagation. It is applicable to space communications and, indeed, this is how light from the sun and other stars reaches the Earth. It is not restricted to lightwaves, though, and can apply at any frequency. This is the form referred to earlier as having $1/r^2$ dependence of power density with distance. Within the Earth's atmosphere, however, attenuation owing to absorption modifies this dependence and makes the range attainable strongly dependent on frequency. Each of the various radio propagation modes is more appropriate at some frequencies than others. However, taken together these different modes make possible a wide variety of radio communication systems.

More detail on radio propagation in telecommunications can be found in Hills, M.T. and Evans, B.G., *Transmission Systems*, George Allen & Unwin, 1973, chapter 5.

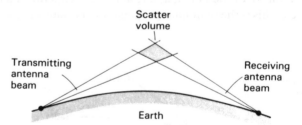

Fig. 1.12   Tropospheric scatter propagation.

Communication Networks

It was observed at the outset that telecommunications involves the sending of messages from a source to a sink. It should not be assumed that this communica-

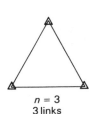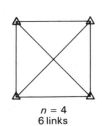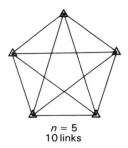

| $n = 3$ | $n = 4$ | $n = 5$ |
| 3 links | 6 links | 10 links |

Fig. 1.13   Examples of fully connected networks.

tion takes place in isolation; frequently there are many possible sources and sinks and messages may need to be sent between varying source–sink combinations. To simplify the discussion, while at the same time increasing its generality, co-located source–sink pairs referred to as *nodes* are considered. A node may thus act as a source, a sink or, in the more general case, simultaneously as a source and a sink. The problem is now one of communication between nodes and the question arises as to how the nodes should best be connected together to form a communication network. It can be seen that there is no single, univerally applicable solution to the network structuring problem; different interconnection patterns and techniques are used depending on circumstances.

Given a set of *n* nodes required to communicate with one another, such that any node may wish to send a message to any other node, we can form a *fully connected* network by introducing a separate bidirectional link between each pair of nodes. This arrangement is illustrated in Fig. 1.13 for certain small-*n* cases from which it is clear that the number of links required increases rather rapidly with *n*. A fully connected *n*-node network has $n(n - 1)/2$ links and the network size–complexity thus increases approximately with the square of the number of nodes.

Exercise 1.2

A computer-communication subsystem contains 8 processing nodes joined together to form a fully connected network. If an extra processor is to be incorporated into the system how many extra links are required to preserve the fully connected structure?

Worked Example 1.3

A local area telephone system has 1000 subscribers. How many extra links would be required to allow a new subscriber to be added if the system employed a fully connected network structure? In view of this, is it likely that telephone networks are actually fully connected?

Solution: The extra telephone instrument would require a separate, direct link to each of the 1000 existing subscribers. Hence, to add just one new subscriber we need 1000 extra links! It seems unlikely, in view of this, that a fully connected structure could economically be employed in a telephone network.

Switched Networks

From the above example it is observed that the cost of adding a new subscriber increases linearly with the number of existing subscribers in a fully connected network. This is a very undesirable characteristic since, if charging for actual

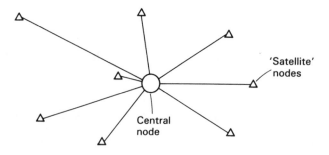

Fig. 1.14 Star network.

installation costs, the 1000th subscriber would pay ten times the line charges of the 100th subscriber. There is a preference for the cost to be both lower and more uniformly distributed. Also, in all probability, most of the links are idle most of the time. A normal telephone system allows for interconnection of pairs of subscribers so that only one link to a subscriber would be in use at any time. Given the excessive line costs and poor link utilization for a fully connected telephone network this structure is not employed. Instead a special central node is introduced which allows lines to be coupled together as required. A star network results, as shown in Fig. 1.14. Now any subscriber can communicate with any other via the central node or *switching centre*. The switching centre (e.g. a telephone exchange) can produce all $n(n-1)/2$ possible connections required but not all at once. If all subscribers were using the system simultaneously then just $n/2$ links would be needed but even this requirement is extremely unlikely to arise. In practise less links are employed and these can be switched between pairs of subscribers as required. The principle is illustrated in Fig. 1.15 where three links are provided to allow a group of 10 subscribers to communicate with one another. The essential feature of this switched network is that expensive items, in this case links, can be time-shared. The statistical properties of the communication requirements of large user groups are relied upon to effect an economy. As a result the amount of new equipment which must be provided when adding a new subscriber is modest and largely independent of the size of the existing group of users.

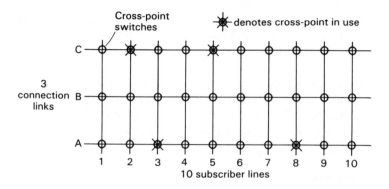

Fig. 1.15 A small communications switching centre. *Note*: subscribers 2 and 5 are communicating via link C; subscribers 3 and 8 are communicating via link A; link B is available.

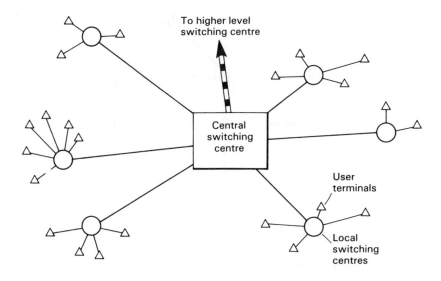

Fig. 1.16 A hierarchical 'star of stars' network.

The idea of a star connection into a switching centre can be extended to form a *star of stars* hierarchical network, as indicated in Fig. 1.16. Users in a restricted geographical area are connected to a local switching centre as described above and local switching centres are connected together via a central switching centre. For all but the minimum of cross-exchange communication traffic (i.e. communication between subscribers on different local exchanges) the links between local and central switching centres must have sufficient communication capacity to allow several messages to be transferred simultaneously. This is achieved by combining the several message waveforms to form a single composite message. This process, known as multiplexing, was mentioned briefly earlier and is discussed in some detail in later chapters.

A large network may have many levels in the hierarchy and it is usual to incorporate additional cross connections, for example between local exchanges, when this is justified by the expected or experienced volume of communication traffic. Expediency demands that practical networks are neither fully connected nor purely hierarchical. This is illustrated by the *transmission plan* of Fig. 1.17 which shows various types of telephone switching centre and possible interconnections.

Of course, the adoption of a switched network configuration carries with it certain overheads. In particular, there is a need for extra equipment to control the network. A user must first communicate with the switching centre his or her wish to make a call together with the identity of the called party. A free connecting link must be identified and the switching centre must then check whether or not the called party is free. If so, the switching centre must communicate with the called party sending a signal to get attention. The link can then be completed and messages transferred. At the end of the conversation the link is returned to the pool to be made available to other users as required. The control of a large switching system is a complex task now frequently delegated to a computer program. This form of switching systems control is referred to as *stored program control* (s.p.c.).

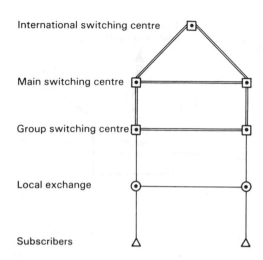

International switching centre

Main switching centre

Group switching centre

Local exchange

Subscribers

Fig. 1.17 A national transmission plan showing cross-links in the hierarchy.

More details on switched networks with particular reference to telephony are provided in Hills, M.T., *Telecommunication Switching Principles*, George Allen & Unwin, 1979.

Data Networks

Our discussion up to this point has been couched in terms of speech telephony but the ideas outlined are more broadly applicable. For example, the special signals used in establishing a telephone connection are an aspect of the *signalling sub-system*. This is an important consideration in the development of any new telecommunications system and has its counterpart in the more general communications context. For data and computer communications the link set-up procedure is referred to as a *link-level protocol*. A protocol is a well-defined procedure to enable terminals/nodes to communicate effectively over a network. Protocols for modern data communication systems can be extremely complex — involving several layers as indicated in Fig. 1.18 — and require careful design and verification.

As a further example of the diverse character of communication networks consider the ring structure of Fig. 1.19a. Any node may communicate with any other by simply applying signals to the ring. Of course, only one message may be transmitted at any time but by time-sharing the circuit between different messages with different source and destination nodes effectively simultaneous transfer of multiple messages is achieved. This is the essence of a form of *local area network* (LAN) used for computer/data communication, developed at the University of Cambridge and known as the *Cambridge Ring*. Data circulate around the ring in *packets*, a packet being a sequence of binary digits represented by a pulse waveform. A hypothetical packet is shown in Fig. 1.19b. Note that a packet may be empty, in which case a node wishing to transmit a message may insert data into the packet, or it may be full. All nodes monitor the packets as they circulate around the ring since the data destination is contained in the *header*. A node reads the data in the packet if it is the intended destination and is free to accept data. It indicates that it has done so by changing the response bits. A packet is cleared on return to the source node. More than one packet may be circulating around the ring allowing several messages to be transferred simultaneously by time-sharing the physical ring connection.

The technique of data communication by way of packets is also used in other

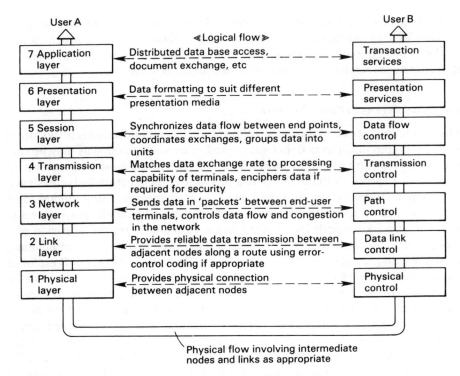

Fig. 1.18   Protocol layering.

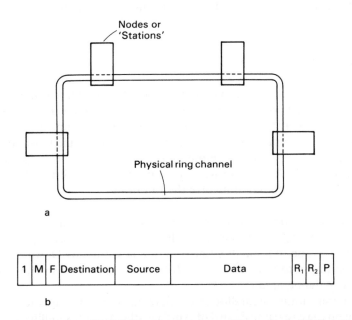

| 1 | M | F | Destination | Source | Data | R₁ | R₂ | P |

b

Fig. 1.19   A ring network. (a) Ring network configuration; (b) Cambridge ring data packet: M = monitor bit, F = full or empty bit, P = parity bit (see Chapter 7), 'destination' and 'source' are 8-bit addresses, R₁ and R₂ are response bits changed by receiving station to indicate action taken.

network structures. The nodes in a network examine the header and send the packet along an appropriate route towards its destination. Packet switching networks of this type may be geographically distributed over a wide area. They are thus sometimes referred to as *wide area networks* (WANs).

Integrated Services Digital Network

The distinction between data and telephone speech communication is by no means sharp. It is seen in Chapter 6 that speech and other analogue signals may be converted into the form of digital data. Indeed, this is now the most common way of transmitting speech in the higher levels of the telephone network. Increasingly, both signal transmission and switching is implemented using digital data techniques and the telephone systems of many countries are rapidly moving over to a digital network. In addition, there is an increasing need for the provision of a wider range of communication services: data transmission, electronic mail, facsimile (still picture) transmission and so on. Given the existence and extent of the telephone network it is not suprising that this is being considered as the basis for the integration of these and many other services into a single network. Means of transmitting data over the telephone network were devised even before digital transmission within the telephone network itself was widespread. More recently there has been a conscious attempt to bring together the various services and make these widely available on a single network, with as much uniformity as practicable. These moves towards *service integration* together with the continued expansion of *digital transmission* and switching have led to the proposal of an *integrated services digital network* (ISDN). Just how the telephone network may/should evolve towards this goal is the subject of much current research and debate.

Telecommunications Worldwide

The telephone network has grown up over the years so that it is now possible to place a call to almost anywhere in the world, usually without the assistance of an operator. The call may be carried by a wide variety of communication media: wire pairs, coaxial cables, optical fibres, radio/microwave links and so on. In addition, several links may be used in combination. For example, if telephoning from London to Los Angeles on the west coast of the United States of America, a hypothetical connection might involve the following:

(a) Direct transmission of analogue speech signal via wire pair to local exchange.
(b) Conversion to digital form and switching to a (time-shared) optical fibre link for transmission to a satellite Earth station at Goonhilly Downs.
(c) Modulation on to a 6 GHz carrier for transmission to a satellite positioned over the Atlantic ocean.
(d) Conversion within the satellite to a 4 GHz carrier for transmission to an Earth station on the eastern seaboard of America (the Atlantic satellites cannot communicate directly with California).
(e) Transmission over landlines, using frequency division multiplexing, to a main switching centre within Los Angeles.

(f) Conversion to a multiplex format involving less channels for transmission over cable to another, local, switching centre.

(g) Conversion to baseband for transmission over a wire-pair to the destination subscriber.

Alternatively, the signal might be sent via an undersea cable to the east coast of America and then over land as above. A user generally has no need to know which route a call takes — all that is required is a good connection. There are, however factors which must be taken into account by the network controllers when routeing calls. For example, time differences around the world give rise to non-synchronous peaks of demand in different locations. When it is breakfast time in New York it is still night time in California. As a result, it is not unknown for telephone calls at busy times between New York and Washington D.C. on the east coast to be routed via, say, Los Angeles when all direct links are occupied. Another factor which must be taken into account is signal delay. For telephone speech routed via radio, surface links or undersea cables this presents no difficulties. The use of a satellite in the chain, however, introduces some 270 ms of one-way delay. This represents a pause of approximately 0.5 s between speaking and receiving the answer. This is perceptible but subjectively acceptable to users. However, if two satellites are included in a chain the attendant delay exceeding one second makes conversation virtually impossible. In view of this, telephone connections from England to eastern Australia are restricted to no more than one satellite segment (a satellite over the Indian ocean), the link being completed via cable. Similarly communication from Europe to the west coast of America makes extensive use of land lines even though combined use of the Indian and Pacific ocean satellites would avoid this.

This constraint does not apply to one-way communication such as broadcast television or non-interactive data transmission. There is thus great flexibility for information transfer around the world, with routeing strategies taking account of varying demand peaks, tolerance of different services to delay and so on, to allow for effective utilization of this worldwide telecommunications network.

Summary

This chapter has provided a general overview of many aspects of telecommunications with some emphasis on the large-scale nature of the systems involved. A message is represented by a signal corresponding to a voltage or current waveform. The signal, which may represent speech, a television picture or data, for example, can be conveyed over long distances using a suitable communication channel such as a coaxial cable, radio link or optical fibre. To render the signal compatible with the available channel some form of signal processing such as modulation or coding may be required while to make effective use of a wide bandwidth channel several signals may be combined and transmitted simultaneously. There is a need for complex interconnection networks to provide for multi-message communication between many different centres, and there is a move towards an integrated services digital network (ISDN) to allow for digital transmission within a single network structure of a wide variety of different types of signals.

Problems

1.1 A step index multimode fibre has a core refractive index of 1.52, a cladding index of 1.51 and a loss of 5 dB/km.
   (i)   From the point of view of pulse spreading what is the maximum fibre length allowable for data transmission at 2 Mbit/s?
   (ii) What would be the loss of a fibre of this length?
   (iii) If 0.5 mW of optical power is launched into the fibre determine the output power level, expressing this both in terms of watts and in terms of dBm.

1.2 A small satellite ground station is to be up-graded by replacing the 2 m diameter reflecting parabolic dish antenna by one of 4 m diameter. How much improvement in received signal strength would you expect this to provide?

1.3 A software engineering laboratory is equipped with 16 microcomputer-based workstations. It is required to interconnect these so that they can communicate with one another and also with a printer and a separate storage system known as a file server. Determine how many 2-way links would be required for a fully connected network. Suggest an alternative network structure which might be more appropriate in this instance.

1.4 The Cambridge ring operates at 10 Mbit/s but there are only 16 data bits per 40 bit data packet. In view of this, estimate the maximum information transfer rate in bit/s.

Signal Representation and Analysis

2

- [ ] To provide a basis for describing signals analytically in the time domain.
- [ ] To show how signals may be scaled and translated in time.
- [ ] To introduce the concept of a signal spectrum and of a signal description in the frequency domain.
- [ ] To explain what is meant by *negative frequency* and a bilateral or *double-sided* spectrum.
- [ ] To show how signals may be described in either domain and to establish the relationship between these descriptions via Fourier analysis.
- [ ] To illustrate the effect of filtering on a signal spectrum.
- [ ] To note briefly that spectral methods can be extended to encompass signals of arbitrary waveshape such as practical communication signals and noise.
- [ ] To introduce the idea of describing signals probabilistically in terms of the amplitude distribution.

It was seen in the previous chapter that communication links involve the transmission of a signal waveform representing the message from a source to a destination. At the destination the received waveform is processed and the message is recovered as nearly as possible — 'as nearly as possible' because the received waveform generally contains two components: an attenuated and delayed version of the original signal from which the message could be recovered exactly, together with an extraneous, unwanted component which interferes with the wanted part and may give rise to error in the recovered message. The unwanted components may be due to signal distortion by the channel, to interference from nearby electrical equipment, or it may simply be electrical noise: apparently random signal variations of thermodynamic or quantum origin. It is not necessary to be concerned with the physical mechanisms giving rise to these random waveforms but it is necessary to assess their implications for the performance of communication systems. In order to do this both the message or signal waveform and the unwanted noise/interference components must be represented mathematically. In this chapter the various ways in which signals may be represented are examined. Before doing this, however, it is appropriate to note that it is neither necessary nor practicable to assess the performance of a system by examining how it will process all possible message waveforms. For speech communication this would involve a consideration of all possible speakers (including, presumably, some not yet born) and all possible sentences (some not yet uttered)! Instead prototype signals with characteristics similar to those of the actual signals to be conveyed are used. For example, for speech telephony much useful analysis can be based on how a system responds to sinusoidal signals of different frequencies and amplitudes, while for data communication the response of a system to a single pulse may provide an insight into how the system responds to a message waveform corresponding to a sequence of

Signal modelling and analysis more generally is discussed in another text in this series: Meade, M.L. and Dillon, C.R., *Signals and Systems: models and behaviour*, Van Nostrand Reinhold, 1986 — which provides a useful complement to the treatment presented here.

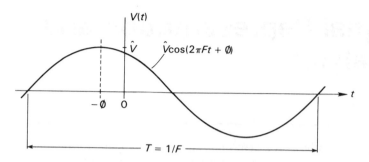

Fig. 2.1   A sinusoidal signal in the time domain.

pulses. In view of this the study of signals is commenced by considering the representation and analysis of such idealized signals.

The Time Domain

A signal waveform may be viewed as the variation with time of a quantity such as voltage or current. For the sake of definiteness consider voltage waveforms. As a first example, consider the sinusoidal voltage signal given by

$$v(t) = \hat{V}\cos(2\pi Ft + \phi) \tag{2.1}$$

Here $\hat{V}$ is the peak voltage, F is the frequency and ϕ is the relative phase. These parameters are indicated on the diagram of Fig. 2.1. The waveform is periodic with period $T = 1/F$ since

$$v(t) = v(t + T) \tag{2.2}$$

Signals may be observed directly in the time domain by using an oscilloscope.

This description is called a representation of the signal in the time domain: the signal is viewed as a function of time. This is by far the most common way of representing signals and of observing them in the laboratory. Whenever an oscilloscope is used to observe a voltage waveform the signal is being viewed in the time domain. It is necessary to be able to describe signals analytically in the time domain; some examples are given below.

Some Examples of Signals in the Time Domain

Consider a rectangular pulse signal, as shown in Fig. 2.2a. The signal waveform is zero up to some time $t = -1/2$ at which point it steps up to a value 1. The waveform value is 1 over the interval $(-1/2, 1/2)$ and at the end of this interval drops down to zero. The waveform value is zero for all $t > 1/2$. Expressing this analytically

$$x_1(t) = \text{rect}(t) \triangleq \begin{cases} 1 & |t| < 1/2 \\ 0 & \text{elsewhere} \end{cases} \tag{2.3}$$

Here $\text{rect}(t)$ is simply a shorthand way of describing a rectangular pulse with unit height and unit width centred on the origin of the time axis. A similar rectangular pulse occurring at some other time, say centred on $t = t_0$ as shown in Fig. 2.2b, may be expressed in terms of the rect( ) function as follows:

24

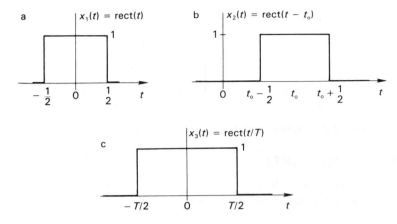

Fig. 2.2   Rectangular pulse signals.

$$x_2(t) = \text{rect}(t - t_0) \tag{2.4}$$

In order to confirm this, note that the centre of the rectangular pulse corresponds to the argument of the function, rect(u) being zero.
Here

$$u = t - t_0$$

and

$$u = 0 \;\rightarrow\; t - t_0 = 0 \;\rightarrow\; t = t_0$$

The pulse is centred on $t = t_0$. This is referred to as signal translation or shifting on the time axis. Notice that subtracting a constant, t_0, from t in the function argument has the effect of shifting the function to the right; that is, to a delay of t_0.

Consider now a pulse of non-unit width, say of duration T, as shown in Fig. 2.2c. This may be expressed analytically as

Note that a minus sign in the argument produces a shift along the positive time axis.

$$x_3(t) = \text{rect}(t/T) \tag{2.5}$$

To check the validity of this formulation note that rect(u) steps from 0 to 1 at $u = -1/2$ and from 1 to 0 at $u = 1/2$. That is, the transitions occur at $|u| = 1/2$. Putting $u = t/T$ it is concluded that transitions occur at $|u| = |t|/T = 1/2 \;\rightarrow\; |t| = T/2$. Transitions occur at $t = \pm T/2$ and the pulse is of width T. This is referred to as signal scaling on the time axis, or simply as *time scaling*.

Note that *dividing* the argument by T has the effect of *multiplying* the pulse width by T.

Amplitude scaling gives no difficulty. A rectangular pulse of height A is simply $A\,\text{rect}(t)$.

Consider a rectangular pulse signal of height A and duration T centred at a point in time $t = t_0 > T$. Sketch the signal waveform in the time domain and obtain an analytic representation in terms of the rect( ) function.

Worked Example 2.1

Solution: Since $t_0 > T/2 > 0$ the pulse is shifted to the right as shown in Fig. 2.3. The signal may be expressed analytically as

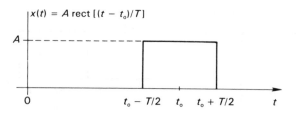

Fig. 2.3   Time-translated rectangular pulse of duration T.

$$x(t) = A \; \text{rect}[(t - t_0)/T]$$

This may be verified as follows:
(i) The rect(u) function is centred on $u = 0$.
Here

$$u = (t - t_0)/T$$

and

$$u = 0 \; \Rightarrow \; (t - t_0)/T = 0 \; \Rightarrow \; t = t_0$$

Hence the pulse of Equation 2.6 is centred on $t = t_0$.

Note carefully this technique. It is useful when sketching signals which have been scaled and/or translated in time.

(ii) Transitions occur in rect(u) when $u = \pm 1/2$.
Here

$$u = (t - t_0)/T$$

and

$$u = \pm 1/2 \; \Rightarrow \; (t - t_0)/T = \pm 1/2$$
$$\Rightarrow \; t - t_0 = \pm T/2 \; \Rightarrow \; t = t_0 \pm T/2$$

Hence transitions are separated by T; the pulse is of width T.
(iii) The amplitude of the pulse corresponds to the value at $t = t_0$.

$$A \; \text{rect}[(t - t_0)/T] \Big|_{t = t_0}$$
$$= A \; \text{rect}(0)$$
$$= A, \text{ since rect}(0) = 1.$$

Consider now how a binary digital signal, such as the pulse sequence of Fig. 2.4a, may be represented. In this example, the signal consists of pulses of width T located at $t = 0, 2T, 3T$ and $5T$. Thus:

$$x(t) = \text{rect}(t/T) + \text{rect}[(t - 2T)/T] + \text{rect}[(t - 3T)/T] + \text{rect}[(t - 5T)/T] \tag{2.6}$$

where each term describes the corresponding pulse. More generally:

$$x(t) = a_0 \; \text{rect}(t/T)$$
$$+ a_1 \; \text{rect}[(t - T)/T]$$

26

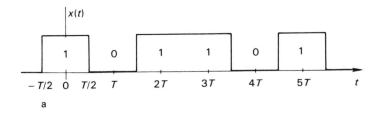

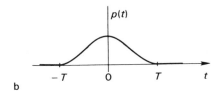

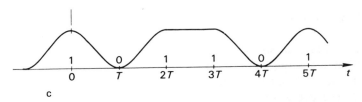

Fig. 2.4   Binary data signals in the time domain. (a) Non-return to zero (NRZ) binary data signal; (b) smoothed pulse signalling element; smoothed data signal based on the signalling element of (b).

$$+ a_2 \, \text{rect}[(t - 2T)/T]$$
$$+ \ldots + a_n \, \text{rect}[(t - nT)/T] + \ldots$$
$$= \sum_n a_n \, \text{rect}[(t - nT)/T] \tag{2.7}$$

The notation $\sum\limits_n$ means 'add up the terms for all relevant integer values of n'.

where $\{a_n\}$ represents the data, a_n taking the value 1 or 0 according to whether or not a pulse is present in the corresponding time slot of duration T centred on $t = nT$. The summation is taken over all relevant (integer) values for n. Very often the signal is assumed to exist throughout all time and can be written

$$x(t) = A \sum_{n=-\infty}^{\infty} a_n \, \text{rect}[(t - nT)/T] \tag{2.8}$$

with $a_n \in \{0, 1\}$ for a unipolar binary data signal of amplitude A. For a bipolar signal $a_n \in \{-1, 1\}$ while for a unipolar, multilevel pulse signal we have $a_n \in \{0, 1, \ldots (N - 1)\}$ if the signal can assume one of N uniformly separated values in each time slot.

A similar procedure may be used to describe pulse sequences in which the basic pulse element, the *signalling element*, waveform is not rectangular. Denoting the elemental pulse waveform as $p(t)$,

$$x(t) = \sum_{n=-\infty}^{\infty} a_n p(t - nT) \tag{2.9}$$

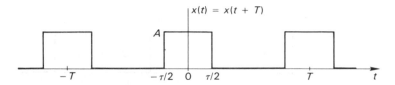

Fig. 2.5   A rectangular wavetrain.

An illustrative segment of such a signal is shown in Fig. 2.4b, c based on a pulse waveform $p(t)$ of raised cosine form:

$$p(t) = \begin{cases} [1 + \cos(\pi t/T)]/2 & |t| < T \\ 0 & \text{elsewhere} \end{cases} \qquad (2.10)$$

Exercise 2.1   Obtain an analytic expression for the periodic rectangular wavetrain of Fig. 2.5.

The Frequency Domain

Consider once more the sinusoidal voltage waveform of Equation 2.1. The word 'sinusoidal' indicates the shape of the waveform. Given a sinusoidal signal a complete specification of the signal is provided by the amplitude $\hat{V}$, the frequency F and the phase ϕ. These three parameters, together with the knowledge that the signal is sinusoidal, are sufficient to be able to draw the voltage waveform; the triple ($\hat{V}$, F, ϕ) fully specifies the signal. For the time being attention is restricted to signals of strictly cosinusoidal form, i.e. $\phi = 0$, and we will deal with just the amplitude $\hat{V}$ and frequency F: now the *pair* ($\hat{V}$, F) fully specifies the signal. In this representation F and $\hat{V}$ can take any positive value and the signal can be represented graphically as shown in Fig. 2.6. This is a representation in the *frequency domain*. With this view, a cosinusoidal signal is represented by a vertical arrow located at some point F on the frequency axis and the height of the arrow corresponds to $\hat{V}$, the amplitude of the signal. Note that if $F = 0$ then there is a d.c. signal with amplitude V. This can also be represented in the frequency domain by an arrow of height V located at the origin.

The frequency domain representation is very useful because more complicated signals can be considered as a superposition (i.e. a summation) of sinusoidal components with different amplitudes, frequencies and phases. For periodic signals the frequencies are integer related. If T is the period then the waveform is said to have a fundamental frequency of $1/T$. The next lowest frequency contained in

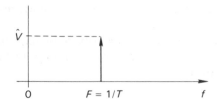

Fig. 2.6   A sinusoidal signal in the frequency domain.

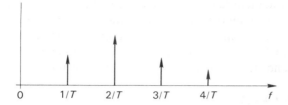

Fig. 2.7   A non-sinusoidal, periodic signal viewed in the frequency domain.

the waveform is $2/T$, termed the second harmonic, and after that comes the third harmonic, frequency $3/T$, and so on. Generally, the various harmonic components have different amplitudes and phases (some may have zero amplitude). To obtain a general appreciation of the harmonic content of a complicated waveform it is convenient to represent the signal graphically in the frequency domain, as illustrated in Fig. 2.7 (once again, phase is neglected). The heights of the arrows represent the strengths of the various harmonic components — the amplitudes of the various cosinusoidal components comprising the complicated periodic waveform. This frequency domain representation is often called the *spectrum* of the signal; the various constituent frequency components are the *spectral components* or *spectral lines*. It was noted previously that it is common practice to view signals in the time domain using an oscilloscope. It is possible also to view signals in the frequency domain (neglecting phase) using an instrument called a *spectrum analyser*. Fig. 2.8 shows an illustrative signal spectrum as displayed on a spectrum analyser.

Signals may be observed in the frequency domain by using a spectrum analyser.

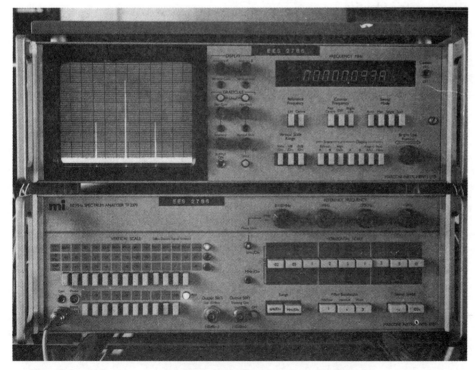

Fig. 2.8   A frequency domain view of a signal as displayed on a spectrum analyser.

The frequency domain representation presented up to this point is incomplete in that it contains only amplitude information and omits any mention of the relative phases of the various components. This can be remedied in various ways. One solution is to label each arrow with the phase of that component. That is, each component corresponds to a time domain contribution of the form

$$A_n \cos(2\pi F_n t + \phi_n)$$

where A_n is the amplitude, $F_n = n/T$ is the frequency and ϕ_n is the phase. A composite signal may be the sum of many terms of this form and may also contain a d.c. component. A periodic signal $x(t)$ with period T can be represented as follows:

$$\begin{aligned}
x(t) = A_0 &+ A_1 \cos[2\pi(t/T) + \phi_1] \\
&+ A_2 \cos[2\pi(2t/T) + \phi_2] \ldots \\
&+ A_n \cos[2\pi(nt/T) + \phi_n] + \ldots
\end{aligned} \tag{2.11}$$

or, more concisely, as

$$= A_0 + \sum_{n=1}^{\infty} A_n \cos[2\pi(nt/T) + \phi_n] \tag{2.12}$$

This is known as a *Fourier series* representation for $x(t)$. Note its close relationship with the frequency domain representation: the amplitudes A_n give the heights of the various spectral lines.

An alternative Fourier series representation may be obtained by using the trigonometric identity

$$\cos(A + B) = \cos(A)\cos(B) - \sin(A)\sin(B) \tag{2.13}$$

to write

$$\begin{aligned}
A_n \cos[2\pi(nt/T) + \phi_n] &= A_n[\cos(2\pi nt/T)\cos(\phi_n) - \sin(2\pi nt/T)\sin(\phi_n)] \\
&= [A_n \cos(\phi_n)]\cos(2\pi nt/T) + [-A_n \sin(\phi_n)]\sin(2\pi nt/T) \\
&= a_n \cos(2\pi nt/T) + b_n \sin(2\pi nt/T)
\end{aligned} \tag{2.14}$$

where

$$a_n = A_n \cos(\phi_n)$$
$$b_n = -A_n \sin(\phi_n)$$

We can thus represent $x(t)$ in the form

$$x(t) = A_0 + \sum_{n=1}^{\infty} a_n \cos(2\pi nt/T) + \sum_{n=1}^{\infty} b_n \sin(2\pi nt/T) \tag{2.15}$$

That is, a periodic signal can be considered as a sum of:
  (i) a d.c. term of amplitude A_0,
  (ii) a set of harmonically related cosinusoidal signals with amplitude a_n, and
  (iii) a set of harmonically related sinusoidal signals with amplitude b_n.
Yet another representation makes use of the observation that

$$\cos A = \frac{\exp(jA) + \exp(-jA)}{2} \tag{2.16}$$

whence

$$A_n \cos[(2\pi nt/T) + \phi_n]$$
$$= \frac{A_n}{2}\left\{\exp[j(2\pi nt/T + \phi_n)] + \exp[-j(2\pi nt/T + \phi_n)]\right\}$$
$$= \frac{A_n}{2}\left\{\exp(j2\pi nt/T)\exp(j\phi_n) + \exp(-j2\pi nt/T)\exp(-j\phi_n)\right\}$$
$$= c_n \exp(j2\pi nt/T) + c_{-n}\exp(-j2\pi nt/T) \qquad (2.17)$$

where

$$c_n = \frac{A_n}{2}\exp(j\phi_n)$$

and

$$c_{-n} = \frac{A_n}{2}\exp(-j\phi_n) = c_n^* \qquad \text{assuming } A_n \text{ is real}$$

The asterisk notation for complex conjugates is being used here: c_n^* is the complex conjugate of c_n.

Thus $x(t)$ can be expressed as

$$x(t) = A_0 + \sum_{n=1}^{\infty}[c_n \exp(j2\pi nt/T) + c_{-n}\exp(-j2\pi nt/T)] \qquad (2.18)$$

Using the fact that

$$\exp(j2\pi nt/T)\big|_{n=0} = e^0 = 1 \text{ for } n = 0$$

and defining $c_0 = A_0$, $x(t)$ can be expressed in terms of a single series ranging over all integers:

$$x(t) = \sum_{n=-\infty}^{\infty} c_n \exp(j2\pi nt/T) \qquad (2.19)$$

This, the exponential form of the Fourier series, is particularly useful in signal analysis. The c_n values are complex numbers with $|c_n|$ corresponding to the magnitudes and $\arg(c_n)$ to the phases of the constituent spectral components. In this representation, however, n takes both positive and negative values and it is convenient to consider a two-sided spectrum of the form shown in Fig. 2.9. By analogy with the earlier discussion of graphical representation of a signal spectrum, each arrow corresponds to a spectral line; the nth term is a component of freqency n/T.

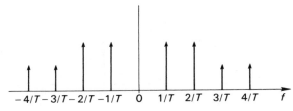

Fig. 2.9 A two-sided (bilateral) frequency domain representation.

31

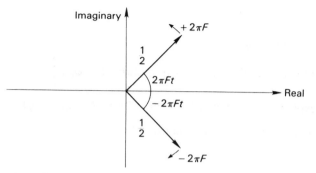

Fig. 2.10   A cosine wave represented by two contra-rotating vectors in an Argand diagram.

The concept of negative frequency described here is extremely important. Take time to become familiar with the idea.

Since n ranges over both positive and negative values this makes it necessary to give some meaning to *negative frequency*. The idea may take a little getting used to! It is not being suggested that a real signal has a negative frequency — nor indeed that it has a positive frequency. A real signal has a *frequency*, such as F Hz. However, a real signal may be considered for the purposes of analysis as the sum of positive and negative frequency components, just as a cosine wave may be considered as the sum of two complex exponential factors.

With this perspective

$$\cos(2\pi Ft) = \text{cosinusoidal signal with frequency } F$$
$$= \frac{\exp(j2\pi Ft) + \exp(-j2\pi Ft)}{2}$$
$$= \frac{1}{2}\exp(+j2\pi Ft) \qquad + \frac{1}{2}\exp(-j2\pi Ft) \qquad (2.20)$$
$$\textit{positive frequency term} \quad \textit{negative frequency term}$$

The signal may be represented on an Argand diagram by two contra-rotating vectors, as shown in Fig. 2.10. The vector rotating in the positive (anticlockwise) direction has angular velocity $+2\pi F$ and is said to correspond to a positive frequency term. The vector rotating in the negative (clockwise) direction has angular velocity $-2\pi F$ and is said to correspond to a negative frequency term. The sum of the two components, corresponding to the signal $x(t)$, is always real. Each individual component is a complex function of time but since both terms always occur together, as complex conjugates, the resultant is always real. Thus the signal $\hat{V}\cos(2\pi Ft)$ can be represented in the frequency domain by a diagram of the form shown in Fig. 2.11. With this double-sided representation a cosine wave has two frequency components of equal strength located at $f = \pm F$. This says no more than

The *bilateral* frequency domain representation may seem unnecessarily cumbersome. At a later stage though it will be found to provide considerable analytic convenience.

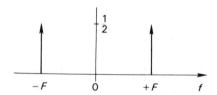

Fig. 2.11   Bilateral frequency domain representation for a cosine wave.

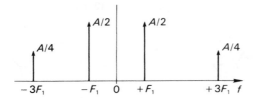

Fig. 2.12   Bilateral spectrum for a signal of Equation 2.21 with $F_2 = 3F_1$ and $B = A/2$.

that we are dealing with a cosine wave of frequency F but are using a model in which this is viewed as a sum of two vectors rotating in opposite directions.

Some Examples of Signals in the Frequency Domain

(i) *Sum of cosine waves*. Consider a signal comprising two cosine waves of different frequencies:

$$x(t) = A \cos(2\pi F_1 t) + B \cos(2\pi F_2 t) \tag{2.21}$$

Expanding this in terms of complex exponential factors, we obtain

$$x(t) = \frac{A}{2}\exp(+j2\pi F_1 t) + \frac{A}{2}\exp(-j2\pi F_1 t) + \frac{B}{2}\exp(+j2\pi F_2 t) + \frac{B}{2}\exp(-j2\pi F_2 t)$$
$$\tag{2.22}$$

positive	*negative*	*positive*	*negative*
frequency	*frequency*	*frequency*	*frequency*
term	*term*	*term*	*term*
at $+F_1$	*at $-F_1$*	*at $+F_2$*	*at $-F_2$*

For example, if $F_2 = 3F_1$ and $B = A/2$ this signal has a frequency spectrum of the form shown in Fig. 2.12.

(ii) *Square wave signal*. A square wave can be viewed as a sum of harmonically related cosinusoidal components. This is illustrated graphically in Fig. 2.13. Starting with a term $\cos(2\pi t/T)$ as the fundamental component, we then add a third harmonic component, $-1/3\cos(2\pi 3t/T)$. This has the effect of slightly depressing the peaks whilst sharpening the transitions. The resultant waveform looks a little more like a square wave than does the original cosine wave. If we now add a fifth harmonic term, $+1/5\cos(2\pi 5t/T)$, a seventh harmonic term, $-1/7\cos(2\pi 7t/T)$, and so on; the resultant approaches closer and closer to a square wave. It can be said that the *partial sum* — that is, the sum up to some $(2n-1)$th harmonic — approximates to a square wave, or that the sum converges to a square wave as $n \to \infty$. The square wave signal may be viewed in the frequency domain as shown in Fig. 2.14. Since the double-sided representation is used the fundamental term at $f = 1/T$ gives rise to spectral lines of strength $1/2$ at $f = \pm 1/T$, and so on.

The procedure just outlined may be termed *Fourier synthesis* in that a close approximation to a square wave has been constructed (or synthesized) by adding together sinusoids having appropriate frequency, amplitude and phase relationships one to another. The inverse of this process, Fourier analysis, is

A computer programme to calculate and plot this partial sum is provided in: Attikiouzel, J., *Pascal for Electronic Engineers*, Van Nostrand Reinhold, 1984 (Worked Example 4.5). This provides a useful means of visualizing the approximation process.

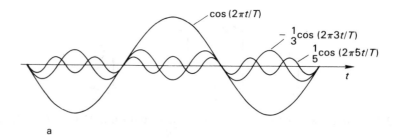

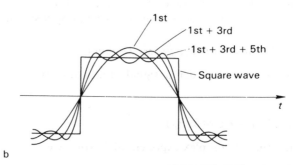

Fig. 2.13   Harmonic synthesis of a square wave. (a) Low-order harmonic
components of a square wave; (b) sequence of partial sums of (a) converge
towards the square wave.

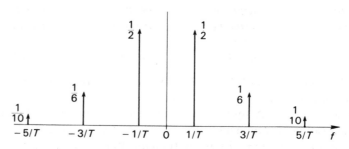

Fig. 2.14   Spectrum of a square wave.

required to determine just what the relationship must be for the partial sum of
harmonically related sinusoids to approach a given signal.

Fourier Series Analysis

It was suggested previously that a periodic signal $x(t) = x(t + T)$ could be
expressed in terms of an exponential Fourier series of the form

$$x(t) = \sum_n c_n \exp(j2\pi nt/T) \tag{2.23}$$

where c_n are the Fourier coefficients corresponding to the amplitudes and phases of
the individual frequency components. These coefficients are related to the time
domain signal by

$$c_n = \frac{1}{T} \int_{-T/2}^{T/2} x(t) \exp(-j2\pi nt/T)\, dt \tag{2.24}$$

We will not formally *prove* this statement but the following argument should provide a sufficiently convincing demonstration that the statement is reasonable.

First note that the choice of indexing integer in Equation 2.23 is arbitrary; thus the following can be written:

$$x(t) = \sum_k c_k \exp(j2\pi kt/T) \tag{2.25}$$

Now substitute Equation 2.25 for $x(t)$ in Equation 2.24:

$$c_n = \frac{1}{T} \int_{-T/2}^{T/2} \left[\sum_k c_k \exp(j2\pi kt/T) \right] \exp(-j2\pi nt/T)\, dt \tag{2.26}$$

where the change from n to k for the indexing integer in Equation 2.25 has avoided a clash with the use of n as an index in Equation 2.24. The strategy now is to demonstrate the consistency of Equation 2.26: to show that the right hand side reduces to c_n. Interchanging the operations of integration and summation:

$$c_n = \frac{1}{T} \sum_k \int_{-T/2}^{T/2} c_k \exp(j2\pi kt/T) \exp(-j2\pi nt/T)\, dt$$

$$= \frac{1}{T} \sum_k c_k \int_{-T/2}^{T/2} \exp(j2\pi(k-n)t/T)\, dt \tag{2.27}$$

Consider now the integral in Equation 2.27 with $m = k - n$:

$$I(m) = \int_{-T/2}^{T/2} \exp(j2\pi mt/T)\, dt$$

$$= \left[\frac{\exp(j2\pi mt/T)}{j2\pi m/T} \right]_{-T/2}^{T/2}$$

$$= \frac{\exp(j\pi m) - \exp(-j\pi m)}{j2\pi m/T} = \frac{\sin(\pi m)}{\pi m/T}$$

$$= \begin{cases} T & m = 0 \\ 0 & m \neq 0 \quad (m \text{ integer}) \end{cases} \tag{2.28}$$

That is, the integral is zero unless $m = k - n = 0$. Hence in Equation 2.27 only the term $k = n$ in the summation is non-zero and

$$c_n = \frac{1}{T} \sum_{\substack{k, \\ k=n}} c_k T = c_n \tag{2.29}$$

Thus it can be concluded that Equations 2.23 and 2.24 are consistent; if a periodic signal $x(t)$ can be expanded according to Equation 2.23, then the coefficients c_n are given by Equation 2.24.

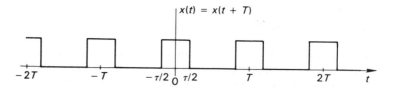

$$x(t) = x(t + T)$$

Fig. 2.15   Periodic rectangular wave.

Worked Example 2.2   Consider a periodic rectangular wave signal as shown in Fig. 2.15 and described by

$$x(t) = \sum_k \text{rect}[(t - kT)/\tau] \tag{2.30}$$

Express this in terms of an exponential Fourier series of the form

$$x(t) = \sum_n c_n \exp(j2\pi nt/T) \tag{2.31}$$

Sketch the signal spectrum, $X(f)$, using the double-sided representation.

Solution: Equation 2.24 gives

$$c_n = \frac{1}{T}\int_{-T/2}^{T/2} \text{rect}(t/\tau) \exp(-j2\pi nt/T)\, dt$$

$$= \frac{1}{T}\left\{\int_{-T/2}^{-\tau/2} 0 \times \exp(-j2\pi nt/T)\, dt + \int_{-\tau/2}^{\tau/2} 1 \times \exp(-j2\pi nt/T)\, dt \right.$$

$$\left. + \int_{\tau/2}^{T/2} 0 \times \exp(-j2\pi nt/T)\, dt \right\}$$

$$= \frac{1}{T}\int_{-\tau/2}^{\tau/2} \exp(-j2\pi nt/T)\, dt$$

$$= \frac{1}{T}\left[\frac{\exp(-j2\pi nt/T)}{-j2\pi n/T}\right]_{-\tau/2}^{\tau/2}$$

$$= \frac{1}{T}\frac{\exp(-j\pi n\tau/T) - \exp(j\pi n\tau/T)}{-j2\pi n/T} = \frac{1}{T}\frac{\sin(\pi n\tau/T)}{\pi n/T}$$

$$= \frac{\tau}{T}\frac{\sin(\pi n\tau/T)}{(\pi n\tau/T)} \triangleq \frac{\tau}{T}\text{sinc}(\pi n\tau/T) \tag{2.32}$$

and the time signal may be expressed as

$$x(t) = \sum_n \frac{\tau}{T}\text{sinc}(\pi n\tau/T) \exp(j2\pi nt/T) \tag{2.33}$$

The signal spectrum thus has the form shown in Fig. 2.16.

With reference to the above example, notice from the spectrum of Fig. 2.16 that spectral lines occur at integer multiples of $1/T$, the fundamental frequency, and

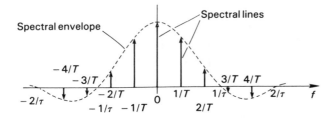

Fig. 2.16   Spectrum of a rectangular wave.

that the strength (amplitude) of these lines is determined by the spectral envelope, which may be defined as

$$E_x(f) = \tau \operatorname{sinc}(\pi f \tau)/T$$

This coincides with Equation 2.33 at $f = n/T$. It can be concluded that the general shape of the spectrum depends on the amplitude and width of the elemental pulse, $\operatorname{rect}(t/T)$, while the line structure is governed by the pulse repetition frequency, $1/T$.

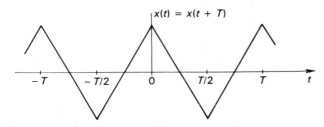

Fig. 2.17   Periodic triangular wave.

Consider the periodic triangular wave signal $x(t)$ shown in Fig. 2.17.
(i) Suggest possible frequencies which may be expected to be identified in this signal if it were subjected to Fourier series analysis.
(ii) Express $x(t)$ as an exponential form Fourier series and calculate non-zero coefficients up to $|n| = 5$. Hence sketch the signal spectrum, $X(f)$, over a corresponding frequency range.

Exercise 2.2

Spectrum of a Train of Narrow Pulses

Consider now a periodic rectangular wave described by

$$x(t) = \sum_n \frac{1}{\tau} \operatorname{rect}[(t - nT)/\tau] \qquad (2.34)$$

Note in this instance the individual pulses have unit area whatever value of $\tau < T$ is adopted. Consider the spectrum as $\tau \to 0$; the sequence illustrated in Fig. 2.18 is obtained. As $\tau \to 0$ the individual spectral components, while preserving their frequency separation of $1/T$, get closer and closer to all having the same strength. This limit case forms a useful approximation when dealing with a train of narrow pulses of large amplitude.

Use is made of this result when sampling is studied in Chapter 5.

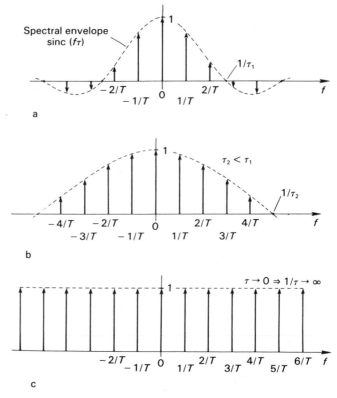

Fig. 2.18 Spectrum of rectangular wave of Equation 2.34, with unit area per pulse, for various values of pulse width τ.

Frequency Domain Representation of Aperiodic Signals

It is possible to obtain a meaningful frequency domain representation for non-periodic signals. However, such signals have continuous spectra since they have no well-defined period and thus do not generally give rise to discrete spectral lines. First consider pulse signals which are represented in the frequency domain by way of a *Fourier integral* or *Fourier transform*. (This is not examined in detail here but it is useful to consider in general terms the frequency domain representation for aperiodic signals.) Recall the periodic rectangular wave of Fig. 2.15 described by Equation 2.30 with spectrum as shown in Fig. 2.16. (For convenience the spectrum is normalized to unit amplitude at $f = 0$.) Letting $T \to \infty$ while keeping τ constant the line components, which are spaced $1/T$ apart, come closer and closer together. In the limit these lines may be thought of as merging together to form a continuous spectrum.

It is possible to obtain a unified treatment encompassing the Fourier series and Fourier integral. Some care is needed, however, concerning convergence and it is generally mathematically more satisfactory to obtain the series from the integral.

As noted previously the envelope or overall shape of the spectrum is determined by the pulse width. Hence it is reasonable to assume that a rectangular pulse of width τ has a continuous spectrum of the form shown in Fig. 2.19a. Notice that the signal is largely concentrated at low frequencies, $|f| < 1/\tau$. We might reasonably expect, therefore, that a lowpass filter would transmit the signal relatively undis-

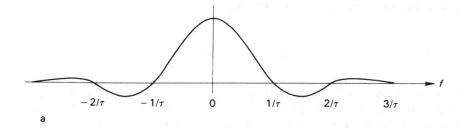

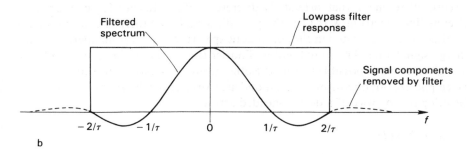

Fig. 2.19   Continuous spectrum and the influence of filtering viewed in the
frequency domain. (a) Continuous spectrum of isolated rectangular pulse;
(b) influence of lowpass filter.

torted if the filter bandwidth B is rather greater than $1/\tau$, as suggested by
Fig. 2.19b. One of the benefits of representing signals in the frequency domain is
that it is rather easy to appreciate the influence of filtering operations on the signal
spectrum.

Fourier Transforms

Consider a signal $x(t)$ satisfying

$$\int_{-\infty}^{\infty} x^2(t)\, dt < \infty \tag{2.35}$$

Such a signal is said to have *finite energy* since if $x(t)$ corresponds to the voltage
across a 1 Ω resistor Equation 2.35 represents the energy dissipated. An example of
a finite energy signal is the rectangular pulse $\text{rect}(t/T)$ while a cosine wave or
square wave provides an example of infinite energy but finite average power signal.
The Fourier series provides a means of obtaining a frequency domain description
of finite average power periodic signals; a frequency domain representation for
finite energy signals is provided by the Fourier transform.
Consider a finite energy signal $x(t)$. Its Fourier transform $X(f)$ is given by

$$X(f) = \int_{-\infty}^{\infty} x(t)\, \exp(-j2\pi ft)\, dt = \mathscr{F}\{x(t)\} \tag{2.36a}$$

Note the use of $\mathscr{F}$ and $\mathscr{F}^{-1}$ as
operators to denote the forward
and inverse Fourier transform
operations.

and $x(t)$ may be obtained from $X(f)$ by way of an inverse transform:

$$x(t) = \int_{-\infty}^{\infty} X(f) \exp(j2\pi ft)\, df = \mathscr{F}^{-1}\{X(f)\} \tag{2.36b}$$

This form of the inverse Fourier transform differs from that given in some texts because we are making use of the frequency variable f rather than angular frequency.

Comparing Equation 2.36a, b with the expressions employed in obtaining a Fourier series expansion of a periodic signal it is noted that Equation 2.36a plays the same role as the expression for the Fourier coefficients while Equation 2.36b corresponds to the expression for a time function in terms of its coefficients. The essential difference is that instead of a discrete set of coefficients $\{c_n\}$ giving rise to a discrete line spectrum $X_d(f)$ for the periodic signal $x_d(t)$ there is a continuum of coefficients encapsulated in the continuous spectrum $X_c(f)$ for an aperiodic finite energy signal $x_c(t)$. The signal and its Fourier transform are intimately linked in that given one it is possible to find the other using the relations of Equation 2.36. To emphasize this one-to-one correspondence we refer to a *Fourier transform pair*, denoting this by way of a double-headed arrow:

A precise and readable account of the unified treatment of Fourier theory is provided by Lighthill, M.J., *An Introduction to Fourier Analysis and Generalised Functions*, Cambridge University Press, 1958.

$$x(t) \Leftrightarrow X(f) \tag{2.37}$$

Many of the results which are derived earlier for Fourier series have their counterparts for the Fourier transform and a unified treatment is both possible, enlightening and practically useful, although we shall not address this in detail here.

Worked Example 2.3 Find the Fourier transform of
(i) a rectangular pulse located at the origin defined by:

$$x(t) \quad = \operatorname{rect}(t/T)$$

(ii) a rectangular pulse delayed by t_0 defined by

$$y(t) \quad = x(t - t_0) = \operatorname{rect}[(t - t_0)/T]$$

Solution: (i) From Equation 2.36a:

$$X(f) = \int_{-\infty}^{\infty} \operatorname{rect}(t/T) \exp(-j2\pi ft)\, dt$$

$$= \int_{-T/2}^{T/2} \exp(-j2\pi ft)\, dt$$

$$= \left[\frac{\exp(-j2\pi ft)}{-j2\pi f} \right]_{-T/2}^{T/2}$$

$$= \frac{\exp(-j\pi fT) - \exp(+j\pi fT)}{-j2\pi f}$$

$$= \frac{1}{\pi f} \frac{\exp(+j\pi fT) - \exp(-j\pi fT)}{2j}$$

$$= T\frac{\sin(\pi fT)}{\pi fT} \triangleq T\operatorname{sinc}(fT) \tag{2.38}$$

Notice that this is in broad agreement with the earlier intuitive argument based on increasing the period of a rectangular wavetrain. The result is illustrated in Fig. 2.20.

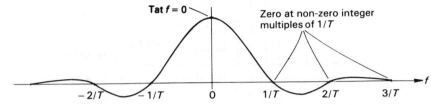

Fig. 2.20   Spectrum of a unit amplitude isolated rectangular pulse.

(ii) Here

$$Y(f) = \int_{-\infty}^{\infty} \text{rect}[(t - t_0)/T] \exp(-j2\pi ft) \, dt$$

$$= X(f) \exp(-j2\pi ft_0) \qquad (2.39)$$

The time delay t_0 simply introduces a phase shift of $-2\pi ft_0$ which is linear with frequency; time translation corresponds to a linear phase shift in the frequency domain.

A Generalized Function and Transform

The Fourier transform is an appropriate tool for dealing with finite energy signals, but it is not restricted to this class of functions. We will consider a very special exception, the *Dirac delta function* $\delta(t)$, defined by:

The Dirac delta function, named after P.A.M. Dirac, the physicist who introduced it in the context of quantum theory, is best thought of as the limit of a sequence of functions as illustrated below:

$$\int_{-\infty}^{\infty} \delta(t) \, dt = 1 \qquad (2.40a)$$

$$\delta(t) = 0, \ t \neq 0 \qquad (2.40b)$$

It is observed that this is a strange 'function' since it is zero everywhere except at $t = 0$ and its value at this point is not defined explicitly. Indeed, this is not a function in the usual sense; it is an example of a *generalized function* and is best defined for our purposes by a limiting operation such as

$$\delta(t) = \lim_{T \to 0} \frac{1}{T} \text{rect}(t/T) \qquad (2.41)$$

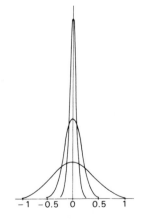

On the right-hand side there is, prior to taking the limit, a rectangular function of width T and height $1/T$. The area is thus unity for all $T > 0$ so Equation 2.40a is satisfied. Now note that $\text{rect}(t/T) = 0$ for all $|t| > T$ so as $T \to 0$ Equation 2.40b is satisfied. With this perspective the delta function is seen as the limiting case of a sequence of rectangular pulses with successively larger amplitudes and smaller widths such that the area of each member function of the sequence is unity. Fig. 2.21 provides an illustration.

Turning to the question of the Fourier transform of $\delta(t)$: the form of the definition of $\delta(t)$ precludes direct computation of the transform. Instead a limiting

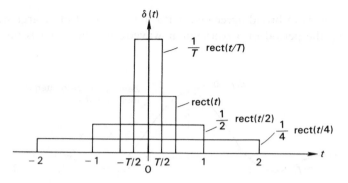

Fig. 2.21 Sequence of unit area rectangular pulses converging to $\delta(t)$ as $T \to 0$.

argument based on Equation 2.41 is employed. For any given $T > 0$ the Fourier transform from Equation 2.38 can be obtained as

$$\frac{1}{T}\text{rect}(t/T) \Leftrightarrow \text{sinc}(fT) \tag{2.42}$$

If $T \to 0$ the $\text{sinc}(fT)$ function gets closer and closer to a constant value of 1, as indicated in Fig. 2.22. A delta function thus has a Fourier transform which is uniform over all frequencies:

$$\delta(t) \Leftrightarrow 1 \tag{2.43}$$

Reciprocity and the Fourier Transform

There is a simple reciprocity theorem which may be stated compactly as

$$\begin{aligned} &\text{If} \quad g(t) \;\Leftrightarrow\; G(f) \\ &\text{then} \quad G(t) \;\Leftrightarrow\; g(-f) \end{aligned} \tag{2.44}$$

We will not concern ourselves with a proof, but note two examples:
(i) *d.c. value.*

$$\begin{aligned} \delta(t) &\Leftrightarrow 1 \\ \Leftrightarrow \quad 1 &\Leftrightarrow \delta(f) \end{aligned} \tag{2.45}$$

The Fourier transform of a d.c. component is represented by a delta function at the origin in the frequency domain.
(ii) *sinc( ) function and the ideal lowpass filter.* Having observed that a

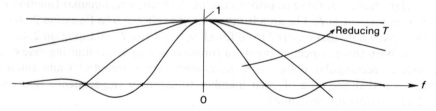

Fig. 2.22 Sequence of sinc (fT) spectra converging to unity as $T \to 0$.

rectangular time domain pulse has a 'sinc function' spectrum, Equation 2.45 gives

$$\text{rect}(t/T) \Leftrightarrow T\text{sinc}(fT)$$
$$\Rightarrow \quad \text{sinc}(t/T) \Leftrightarrow T\text{rect}(fT) \tag{2.46}$$

The function $\text{rect}(fT)$ is an ideal lowpass function in that it is uniform for $|f| < 1/2T$ and zero for $|f| > 1/2T$. It is a useful approximate model for a lowpass filter.

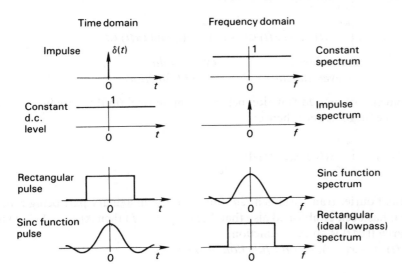

Fig. 2.23   Illustration of reciprocity for Fourier transform pairs.

Symmetry Relationships

Since we are concerned with the analysis of signals corresponding to the variation with time of, for example, a voltage we shall concern ourselves with functions $x(t)$ which are *real* time functions. With this constraint on $x(t)$ the Fourier spectrum $X(f)$ exhibits a special form of symmetry, as follows:

$$
\begin{aligned}
X(f) &= \int_{-\infty}^{\infty} x(t) \exp(-j2\pi ft)\, dt \\
&= \int_{-\infty}^{\infty} x(t) \cos(2\pi ft)\, dt - j \int_{-\infty}^{\infty} x(t) \sin(2\pi ft)\, dt \\
&= \text{Re}[X(f)] + j\,\text{Im}[X(f)]
\end{aligned}
$$

$$
\begin{aligned}
X(-f) &= \int_{-\infty}^{\infty} x(t) \exp[-j2\pi(-f)t]\, dt \\
&= \int_{-\infty}^{\infty} x(t) \cos(2\pi ft)\, dt + j \int_{-\infty}^{\infty} x(t) \sin(2\pi ft)\, dt \\
&= \text{Re}[X(f)] - j\,\text{Im}[X(f)]
\end{aligned}
$$

whence

$$
\begin{aligned}
X^*(-f) &= \text{Re}[X(f)] + j\,\text{Im}[X(f)] \\
&= X(f) \tag{2.47}
\end{aligned}
$$

Hence for a real time signal $x(t)$ the Fourier transform $X(f)$ exhibits *Hermitian symmetry* such that $X(f) = X^*(-f)$.

If additional constraints on the signal are considered then further relationships emerge:

(i) $x(t) = x(-t)$, *an even time function*

$$X(f) = \int_{-\infty}^{\infty} x(t) \exp(-j2\pi ft)\, dt$$

$$= \underbrace{\int_{-\infty}^{\infty} x(t) \cos(2\pi ft)\, dt}_{\substack{even \times even \\ = even\ function}} - j \underbrace{\int_{-\infty}^{\infty} x(t) \sin(2\pi ft)\, dt}_{\substack{even \times odd \\ = odd\ function}}$$

The integral of an odd function between symmetrical (in this case infinite) limits about the origin is zero, hence:

$$X(f) = \int_{-\infty}^{\infty} x(t) \cos(2\pi ft)\, dt \tag{2.48}$$

and the Fourier transform is a *real* function, the imaginary part being zero. But since it has been established also that $X(f) = X^*(-f)$ then $X(f) = X(-f)$ is *real and even* if $x(t)$ is an even function.

(ii) $x(t) = -x(-t)$, *an odd function of time*

$$X(f) = \int_{-\infty}^{\infty} x(t) \exp(-j2\pi ft)\, dt$$

$$= \underbrace{\int_{-\infty}^{\infty} x(t) \cos(2\pi ft)\, dt}_{\substack{odd \times even \\ = odd\ function}} - j \underbrace{\int_{-\infty}^{\infty} x(t) \sin(2\pi ft)\, dt}_{\substack{odd \times odd \\ = even\ function}}$$

This time the real part is zero and $X(f)$ is purely imaginary:

$$X(f) = -j\int_{-\infty}^{\infty} x(t) \sin(2\pi ft)\, dt \tag{2.49}$$

Filtering and the Fourier Transform

The ideal lowpass filter is often expressed in terms of a transfer function of the form

$$H(f) = \text{rect}[f/(2B)] \tag{2.50}$$

which is unity for $|f| < B$ and zero for $|f| > B$. If a signal $x(t) \Leftrightarrow X(f)$ is applied to the input of a lowpass filter the spectrum of the output signal $y(t) \Leftrightarrow Y(f)$ is given by

$$Y(f) = H(f)X(f) \tag{2.51}$$

For an ideal lowpass filter, then,

$$Y(f) = \text{rect}[f/(2B)]X(f)$$

$$\Rightarrow\ Y(f) = \begin{cases} X(f) & |f| < B \\ 0 & \text{elsewhere} \end{cases} \tag{2.52}$$

As observed earlier, one of the features of frequency domain analysis is that it is particularly easy to accommodate filtering operations: filtering a signal corresponds to multiplying the signal spectrum by the filter transfer function. This observation applies equally to periodic and aperiodic signals, the latter being characterized in the frequency domain by way of the Fourier transform.

Impulse Response of a Filter

If we now consider the special case of an input signal spectrum $X(f) = 1$ then from Equation 2.51 the output signal spectrum is $Y(f) = H(f) \times 1 = H(f)$. But $X(f) = 1 \Leftrightarrow x(t) = \delta(t)$, a unit impulse function. Hence the response of a filter to a unit impulse located at $t = 0$ has Fourier transform $H(f)$, the frequency response of the filter. The corresponding time domain signal $h(t) \Leftrightarrow H(f)$ is referred to as the *impulse response* of the filter.

Convolution

We have seen that filtering corresponds in the frequency domain to multiplication of the input spectrum $X(f)$ by the filter transfer function $H(f)$, producing the output signal spectrum $Y(f) = H(f)X(f)$. But to what does this correspond in the time domain? As we shall see, an operation known as *convolution* can be defined which specifies how $x(t)$ and $h(t)$ are combined to produce the output time domain signal $y(t)$. As a useful shorthand notation we let $*$ denote the convolution operation:

$$y(t) = x(t) * h(t) \tag{2.53}$$

defined formally by

$$y(t) = \int_{-\infty}^{\infty} x(u)h(t - u)\,du \tag{2.54}$$

To see how this relates to Equation 2.51 we take the Fourier transform of both sides of Equation 2.54 to obtain

$$Y(f) = \int_{-\infty}^{\infty} y(t)\exp(-j2\pi ft)\,dt$$

$$= \int_{-\infty}^{\infty} \left[\int_{-\infty}^{\infty} x(u)h(t - u)\,du \right] \exp(-j2\pi ft)\,dt$$

$$= \iint_{-\infty}^{+\infty} x(u)h(t - u)\exp(-j2\pi ft)\,du\,dt$$

Interchanging the order of integration allows us to write this as

$$Y(f) = \int_{-\infty}^{\infty} x(u) \left[\int_{-\infty}^{\infty} h(t - u) \exp(-j2\pi ft) \, dt \right] du$$

and using the translation property

$$h(t) \quad \Leftrightarrow \quad H(f)$$
$$\Rightarrow \quad h(t - u) \Leftrightarrow H(f) \exp(-j2\pi fu)$$

This property was illustrated in Worked Example 2.3; time translation corresponds to the introduction of a linear phase shift in the frequency domain.

we obtain

$$Y(f) = \int_{-\infty}^{\infty} x(u)[H(f) \exp(-j2\pi fu)] \, du$$

$$= H(f) \int_{-\infty}^{\infty} x(u) \exp(-j2\pi fu) \, du$$

$$= H(f)X(f) = X(f)H(f)$$

so that

$$y(t) = x(t) * h(t) \Leftrightarrow Y(f) = X(f)H(f) \tag{2.55}$$

The foregoing development has been couched in terms of signal filtering but applies much more generally. Specifically, from the reciprocity relationship for Fourier transforms, Equation 2.44, we note that

$$y(t) = x_1(t)x_2(t) \Leftrightarrow Y(f) = X_1(f) * X_2(f) \tag{2.56}$$

These results may be usefully stated formally as a theorem:

The Convolution Theorem

If two signals are multiplied together in one domain their corresponding (forward or inverse) transforms are convolved in the other domain:

$$x_1(t) * x_2(t) \Leftrightarrow X_1(f) X_2(f) \tag{2.57a}$$
$$x_1(t) x_2(t) \Leftrightarrow X_1(f) * X_2(f) \tag{2.57b}$$

Convolution with Impulses

A system with impulse response $h(t)$ produces as the output signal the waveform $h(t)$ in response to a Dirac delta function applied at the input at time $t = 0$. That is, input $\delta(t)$ yields response $h(t)$. If the input is a delta function at $t = \tau$, i.e. it is $\delta(t - \tau)$, then the output is $h(t - \tau)$ and

$$h(t - \tau) = h(t) * \delta(t - \tau)$$
$$= \delta(t - \tau) * h(t) \tag{2.58}$$

We deduce from this that convolving a pulse $h(t)$ located near $t = 0$ with a delta function located at $t = \tau$ has the effect of producing a time-translated copy of $h(t)$ relocated at $t = \tau$, namely $h(t - \tau)$. Once again this result applies more generally, both in the time domain and in the frequency domain.

For the time domain signal $y(t) = x(t) \cos(2\pi Ft)$ find the Fourier transform $Y(f)$, expressing this in terms of $X(f)$, the Fourier transform of $x(t)$.

Solution: The Fourier transform of the cosine wave $\cos(2\pi Ft)$ corresponds to two impulses in the frequency domain, each of weight $1/2$ located at $f = \pm F$, in accord with the representation of Fig. 2.11. From the convolution theorem if two signals are *multiplied* together in the time domain their Fourier transforms are *convolved* in the frequency domain.

Multiplication in the time domain gives rise to convolution in the frequency domain.

$$\mathcal{F}\{\cos(2\pi Ft)\} = \tfrac{1}{2}\delta(f \pm F)$$
$$\mathcal{F}\{x(t)\} = X(f)$$

Hence

$$\mathcal{F}\{x(t)\cos(2\pi Ft)\} = X(f) * \tfrac{1}{2}\delta(f \pm F)$$
$$= \tfrac{1}{2}X(f \pm F) \tag{2.59}$$

We obtain two copies of the spectrum $X(f)$, scaled in amplitude by $1/2$ and translated in frequency to be centred on $f = \pm F$.

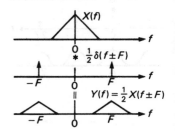

Interpreting Convolution

While Equation 2.54 provides an unequivocal *mathematical* expression for the convolution operation it may be helpful if we also provide a heuristic interpretation. To effect this we shall consider a system with impulse response $h(t)$ driven by an input pulse signal $x(t)$ and producing an output response $y(t) = x(t) * h(t)$.

Let $x(t)$ be approximately represented by a sequence of weighted impulses:

$$x(t) \simeq \tilde{x}(t) = \sum_n x_n \delta(t - n\tau) \tag{2.60a}$$

with

$$x_n = \tau x(n\tau) \tag{2.60b}$$

If the input is a set of weighted, time shifted impulse functions then the output is the summation of a set of weighted (amplitude scaled) time shifted copies of the impulse response. The response of the system to an input $\tilde{x}(t)$, which we shall denote as $\tilde{y}(t)$, is thus given by

$$\tilde{y}(t) = \tilde{x}(t) * h(t)$$
$$= \sum_n x_n h(t - n\tau)$$
$$= \sum_n x(n\tau)h(t - n\tau)\tau$$

Considering now reducing τ, in the limit of $\tau \to 0$ we can associate $n\tau$ with a *continuous* variable u while $\tau = (n - 1)\tau - n\tau$ now represents an elemental increment du and the discrete sum becomes an integral:

$$\tilde{y}(t) \underset{\lim \tau \to 0}{\to} \int x(u)h(t-u)\,du = x(t) * h(t) = y(t)$$

Fourier Transforming Periodic Signals

The convolution theorem provides a useful means of obtaining a Fourier transform description for periodic signals. We can describe a periodic signal $x(t)$ with period T as the convolution of a pulse-like signal $x_T(t)$ with an infinite train of impulses spaced in time by T.

A periodic signal $x(t)$ with period T satisfies $x(t) = x(t+T)$ for all t.

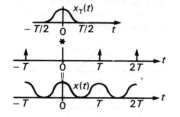

$$x(t) = x_T(t) * \sum_n \delta(t - nT) \qquad (2.61a)$$

where

$$x_T(t) = \begin{cases} x(t) & \text{for } |t| < T/2 \\ 0 & \text{elsewhere} \end{cases} \qquad (2.61b)$$

It follows that the Fourier transform of $x(t)$ is given by

$$X(f) = \mathscr{F}\left\{ x_T(t) * \sum_n \delta(t - nT) \right\}$$

$$= X_T(f)\mathscr{F}\left\{ \sum_n \delta(t - nT) \right\}$$

where

$$X_T(f) = \mathscr{F}\{x_T(t)\} = \int_{-\infty}^{\infty} x_T(t)\exp(-j2\pi ft)\,dt$$

$$= \int_{-T/2}^{T/2} x(t)\exp(-j2\pi ft)\,dt$$

and

The spectrum of a train of delta functions was illustrated in Fig. 2.18 as the limiting case when considering a train of narrow pulses. The weighting factor of $1/T$ in Equation 2.62 arises because we are concerned with a Fourier transform description.

This is the general case for the specific result obtained in Worked Example 2.2.

$$\mathscr{F}\left\{ \sum_n \delta(t - nT) \right\} = \frac{1}{T}\sum_n \delta\left(f - \frac{n}{T}\right) \qquad (2.62)$$

Hence

$$X(f) = X_T(f)\sum_n \delta\left(f - \frac{n}{T}\right) \qquad (2.63)$$

The spectrum consists of impulses (discrete lines) at multiples of the fundamental frequency $1/T$ with weights determined by $X_T(f)$, the transform of the related signal $x_T(t)$.

Frequency Domain Representation for Signals of Arbitrary Waveshape

The Fourier series and integral are powerful tools in signal analysis but, in general, frequency domain representations are required for arbitrary signals. The foregoing

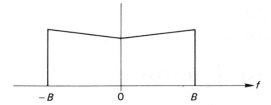

Fig. 2.24 Bandlimited spectrum of an arbitrary lowpass signal.

should make it seem reasonable that a relatively arbitrary signal may have a mean-ingful representation in the frequency domain. In particular, many signals have a lowpass spectrum, being concentrated at low frequencies $|f| < B$ where B is the signal bandwidth. For example, audio signals are concentrated in the band $|f| < 20$ kHz and even if the range of frequencies is restricted to $|f| < 3.4$ kHz by using a lowpass filter it is found that speech is still perfectly intelligible. This latter observation is exploited in telephony. Television signals, on the other hand, while still being essentially 'lowpass', occupy frequencies up to ~5.5 MHz. It is found to be convenient to represent such signals diagrammatically as illustrated in Fig. 2.24. Note, however, that it is not being suggested that the signal spectrum has this precise shape — only that it is of lowpass form concentrated in $|f| < B$.

There is a profoundly important point hidden in the above discussion, and it concerns the nature of the spectral characterization we can hope to obtain for information signals. It was found previously when dealing with deterministic signals that knowledge of the *spectrum* implied knowledge of the time domain signal. We stated this explicitly in terms of the Fourier transform pair; $x(t) \Leftrightarrow X(f)$, but we observed a similar one-to-one correspondence for Fourier series. How, then, can a spectral representation for an arbitrary information signal waveform such as a speech signal be obtained? If the spectrum were known it would be possible to reconstruct the time signal for all values of t. The sort of signals being considered have some degree of randomness associated with them, their precise future behaviour is not known. And yet certain characteristics of the signal, such as its mean power, may be known. For such a signal $x(t)$ the appro-priate spectral characterization is the *power spectrum* or *power spectral density function*, $S_x(f)$. This indicates the concentration of signal power as a function of frequency. Power is a non-negative quantity so $S_x(f)$ is a non-negative function; its integral over all frequencies is the total average signal power:

$$S_x(f) > 0 \text{ for all } f$$

$$P = \int_{-\infty}^{\infty} S_x(f) \, df \tag{2.64}$$

The influence of filtering on signals described by their power spectrum is of interest. Noting that a filter transfer function corresponds to a ratio of signal *amplitudes* and that power is proportional to (amplitude)^2, we conclude that the power spectrum at the output of a filter with transfer function $H(f)$ is given by

$$S_0(f) = S_i|H(f)|^2$$

and the total power is thus

$$P_0 = \int_{-\infty}^{\infty} S_0(f) \, df = \int_{-\infty}^{\infty} S_i(f)|H(f)|^2 \, df \tag{2.65}$$

Amplitude Distribution of Signals

For a deterministic signal such as a cosine wave the time domain description is a precise characterization of the instantaneous variation of the amplitude of the signal; it indicates the amplitude of the signal for each value of the time variable t. For a randomly varying signal such as a speech waveform no such precise and complete description is possible. Nevertheless, it is generally useful (and often necessary) to have some sort of signal characterization relating to possible amplitude values. This is provided by the *amplitude distribution* which is a measure of the relative frequency of occurrence of the various instantaneous values of the signal. The amplitude distribution of a signal $x(t)$ is denoted by

$$p(x) = \lim_{dx \to 0} \frac{\text{fraction of time spent in } dx}{dx} \tag{2.66}$$

where dx is an elemental interval centred on x. An illustrative signal waveform and its corresponding amplitude distribution is shown in Fig. 2.25.

Given the amplitue distribution $p(x)$, the fraction of time the amplitude of a signal $x(t)$ lies within the interval (x_1, x_2) can be calculated as

$$F(x_1 < x < x_2) = \int_{x_1}^{x_2} p(x) \, dx \tag{2.67}$$

Note that the fraction of time a signal spends in an interval can never be negative and that the signal always has some finite amplitude, hence:

$$p(x) \geq 0 \text{ for all } x$$

$$\int_{-\infty}^{\infty} p(x) \, dx = 1 \tag{2.68}$$

An alternative interpretation of $F(x_1 < x < x_2)$ is that it measures the *probability* that a randomly selected sample of the signal $x(t)$ falls within the interval (x_1, x_2).

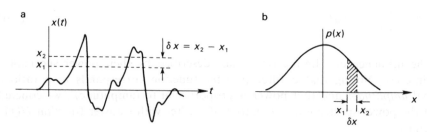

Fig. 2.25   (a) Continuous signal waveform and (b) amplitude distribution.

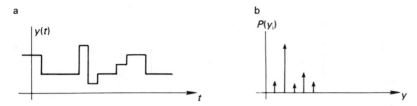

Fig. 2.26   (a) Discrete signal waveform and (b) amplitude distribution.

This probability can be obtained by integrating $p(x)$ over the interval and $p(x)$ is thus usually referred to as a *probability density function*.

The above description relates to continuous signals. Some modification is required for discontinuous signals which take on only certain discrete values. The amplitude distribution is then composed of discrete lines with weights corresponding to the fraction of time the signal takes on the corresponding amplitude value. Fig. 2.26 provides an illustration. For the discrete case, $P(y_i)$ = fraction of time the signal spends at y_i and

$$\sum_i P(y_i) = 1 \tag{2.69}$$

Here the alternative *probabilistic* perspective is that $P(y_i)$ denotes the probability that the amplitude of a random sample of the signal waveform will be *precisely* y_i.

The preceding discussion of amplitude distributions is couched in terms of random signals but they can be applied to deterministic signals too. For example, the amplitude distributions for square, sawtooth and sine waves are illustrated in Fig. 2.27. The square wave takes on just two possible values and thus has a discrete distribution while the triangular and sine wave take on a continuum of values between $-A$ and $+A$ and so have continuous amplitude distributions.

Signal Power

It has been shown previously that the average power in a signal may be obtained by integrating the power spectral density function over all frequencies; it can now be seen how it may be obtained from the probability density function. Assume for convenience that $x(t)$ is a voltage waveform across a 1 Ω resistor. The *instantaneous power* dissipated in the resistor is $x^2(t)$ and the average power is found by averaging over all time:

$$P = \overline{x^2} \triangleq \lim_{T \to \infty} 1/T \int_{-T/2}^{T/2} x^2(t)\ \mathrm{d}t \tag{2.70}$$

Alternatively the average power may be calculated as

$$P = \int_{-\infty}^{\infty} x^2 p(x)\ \mathrm{d}x \tag{2.71}$$

Equation 2.70 is referred to as a time-average and Equation 2.71 as a statistical average. The equivalence of these two expressions is not immediately apparent but may be appreciated as follows: Equation 2.70 indicates that power is the average of

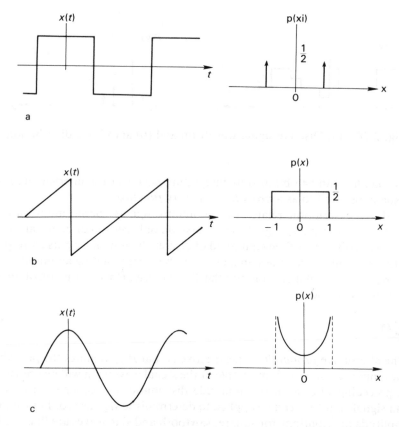

Fig. 2.27   Illustrative waveform (left) and amplitude distribution (right).
(a) Square wave; (b) sawtooth wave; (c) sinusoidal wave.

the square of the signal amplitude; in this instance a time-average is involved. On the other hand, the amplitude distribution measures the relative frequency with which the signal assumes the possible amplitude values and this is based on average behaviour over a long time interval. Hence, if the squares of all possible amplitude values are considered and added up weighted by their relative frequency of occurrence the result is equivalent to having calculated the time-average of the square of the signal. A specific example may help here: consider the sawtooth waveform of Fig. 2.27b. The signal has a well-defined period so the power can be calculated on the basis of a time-average without recourse to the limiting operation as

$$
\begin{aligned}
P &= \frac{1}{T} \int_{-T/2}^{T/2} x^2(t) \ \mathrm{d}t \\
&= \frac{1}{T} \int_{-T/2}^{T/2} \left(\frac{Vt}{T}\right)^2 \ \mathrm{d}t \\
&= \frac{1}{T} \left(\frac{V}{T}\right)^2 \left[\frac{t^3}{3}\right]_{-T/2}^{T/2} \\
&= \frac{1}{T} \left(\frac{V}{T}\right)^2 \frac{T^3}{12} = V^2/12
\end{aligned}
\tag{2.72}
$$

Consider now the statistical approach: the signal amplitude is uniformly distributed on $(-V/2, V/2)$ and the probability density function is thus

$$P(x) = \frac{1}{V} \text{rect}(x/V) \tag{2.73}$$

Equation 2.71 gives the signal power as

$$
\begin{aligned}
P &= \int_{-\infty}^{\infty} x^2 p(x) \, dx \\
&= \int_{-\infty}^{\infty} x^2 \frac{1}{V} \text{rect}(x/V) \, dx \\
&= \frac{1}{V} \int_{-V/2}^{V/2} x^2 \, dx \\
&= \frac{1}{V} \left[\frac{x^3}{3} \right]_{-V/2}^{V/2} \\
&= V^2/12 \text{ as above.}
\end{aligned}
$$

Noise Processes

Communication signals are often corrupted by extraneous, unwanted signals which we term noise, distortion and/or interference. Electrical noise is, in fact, inherent in electrical communications being associated with the random motion due to thermal agitation of charge-carrying electrons.

Here we shall concern ourselves with the spectral characterization of noise. A commonly encountered model treats the noise power as uniformly distributed over all frequencies of interest. The noise is then characterized by its power spectral density η W/Hz so that the power in a band of frequencies from F_1 to F_2 is given by

Noise of thermal origin is referred to as thermal noise. The intrinsic thermal noise power for a system with bandwidth B is given by $P_{\text{noise}} = kTB$ where k is Boltzmann's constant and T is the absolute temperature. Thermal noise is often referred to as kTB (kay-tee-bee) noise.

$$P_N = \int_{F_1}^{F_2} \eta \, df = \eta B \tag{2.74}$$

where B is the bandwidth of interest: $B = F_2 - F_1$. This is referred to as white noise since power is uniformly distributed with frequency.

White noise is so termed by analogy with white light which contains all colours.

Alternatively we may consider a double-sided (bilateral) representation — positive and negative frequencies being separately delimited — in which case the noise power spectral density is $\eta/2$. If this is applied to a filter with transfer function $H(f)$ then the output noise power spectral density is given by

$$S_{N_o}(f) = \frac{\eta}{2} |H(f)|^2 \tag{2.75}$$

Once again we are concerned with the 'power' transfer function $|H(f)|^2$.

If we consider a noise voltage $v_N(t)$ appearing across a 1 Ω resistor then the mean square noise voltage is numerically equal to the noise power. Hence if a noise process has *probability density function* $p_N(v)$ and *power spectral density function*

$S_N(f)$ then the noise power may be obtained as

$$P_N = \int_{-\infty}^{\infty} v^2 \, p_N(v) \, dv \qquad (2.76)$$

using Equation 2.71, or as

$$P_N = \int S_N(f) \, df \qquad (2.77)$$

using Equation 2.64. That is, we may calculate the average noise power on the basis of the amplitude distribution or on the basis of how the noise power is distributed in the frequency domain. The latter is particularly useful when concerned with the influence of filtering on noise.

Worked Example 2.5 Consider a low pass signal $m(t)$ with spectrum $M(f)$ occupying the range of frequencies $|f| < W$ and having signal power $P_S \equiv \overline{m^2}$. This is corrupted by additive white noise. Both signal and noise are to be processed by a lowpass filter with characteristic as shown in Fig. 2.28. Determine the ratio of the signal power to the noise power (the signal to noise ratio, SNR) at the output of the filter and compare this with the result which would be obtained by using an ideal lowpass filter with bandwidth W.

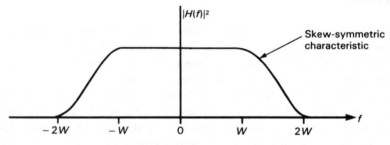

Fig. 2.28   Non-ideal lowpass filter.

Solution: Signal power at output: $P_S = \overline{m^2}$
Noise power at output for filter of Fig. 2.28:

$$P_N = \frac{1}{2} \frac{\eta}{2} W \qquad + \qquad \frac{\eta}{2} 2W \qquad + \qquad \frac{1}{2} \frac{\eta}{2} W$$

$$\left\{ \begin{array}{l} \text{for noise} \\ \text{occupying} \\ -2W<f<-W \end{array} \right\} \qquad \left\{ \begin{array}{l} \text{for noise} \\ \text{occupying} \\ -W<f<W \end{array} \right\} \qquad \left\{ \begin{array}{l} \text{for noise} \\ \text{occupying} \\ W<f<W \end{array} \right\}$$

$$= \tfrac{1}{4}\eta W \qquad + \qquad \eta W \qquad + \qquad \tfrac{1}{4}\eta W$$
$$= 1.5\eta W$$

Hence

$$\text{SNR} = \frac{\overline{m^2}}{1.5\eta W} \qquad (2.78)$$

With an ideal lowpass filter only noise in the band $-W$ to W would be present at the output, with total power ηW, giving

$$\text{SNR}_{\text{ideal}} = \frac{\overline{m^2}}{\eta W} \tag{2.79}$$

whence

$$\frac{\text{SNR}}{\text{SNR}_{\text{ideal}}} = \frac{1}{1.5} \equiv -10 \log_{10}(1.5) = -1.77 \text{ dB} \tag{2.80}$$

Using the filter of Fig. 2.28 thus gives rise to a signal to noise ratio degradation of 1.77 dB compared with using an ideal lowpass filter with bandwidth W.

Summary

In this chapter we have seen how various signals may be described analytically and how a complicated signal can often usefully be viewed as a combination of a large number of simpler signals. We introduced the concepts of the time domain and frequency domain and showed how Fourier analysis provided a link between these two domains. In particular, the Fourier series was introduced and the exponential form emphasized. The idea of 'negative' and 'positive' frequency was discussed and graphical representations for signals in the frequency domain, ranging over negative and positive frequency, were introduced. The role of the Fourier transform in providing a spectral description for both aperiodic and periodic signals was noted.

A summary of some key Fourier transform results is provided in Appendix B.

The amplitude distribution was introduced as another useful characterization for a signal. Distributions were presented and we saw how the signal power may be calculated either as a time average in terms of the signal waveform or as a statistical average in terms of the signal amplitude distribution or probability density function. Finally noise processes were introduced and noise filtering was discussed briefly.

Problems

2.1 Consider the signal element $p(t)$ shown in Fig. 2.4b. This is used as the basis for binary data transmission with data 1s represented by a pulse and 0s by the absence of a pulse. Write down analytic expressions for the signal waveforms corresponding to the data sequence.

$$1\ 0\ 1\ 0\ 1\ 1\ 0\ 1$$

if the data rate is (i) $1/T$ bit/s, and (ii) $1/2T$ bit/s.

2.2 Consider the periodic signal $x(t) = x(t + T)$ defined by

$$x(t) = \sum_{n=-\infty}^{\infty} \text{rect}[3(t - nT)/T]$$

(i) Sketch a representative segment of this waveform in the time domain.
(ii) Express $x(t)$ in terms of an exponential Fourier series and determine values for the coefficients c_n *for* $|n| < 10$.
(iii) Hence sketch $X(f) \Leftrightarrow x(t)$, showing clearly both the form of the spectral envelope and significant detailed structure.

2.3　Let $y(t) = x(t) - 1/3$ where $x(t)$ is as given in Problem 2.2. How does the spectrum $Y(f)$ differ from $X(f)$?

2.4　Consider an infinite train of unit impulses uniformly spaced in time by intervals T.
　　(i)　Sketch this signal in the time domain and write down a corresponding analytic expression.
　　(ii)　Deduce and sketch the form of the spectrum.
　　(iii)　Write down an analytic expression for the spectrum.
　　[*Hint*: Consider the spectrum of a train of narrow unit area pulses as the pulse width is reduced.]

2.5　Consider an isolated triangular pulse defined by

$$x(t) = \begin{cases} 1 - |t|/T & |t| < T \\ 0 & \text{elsewhere.} \end{cases}$$

　　(i)　Sketch this signal in the time domain.
　　(ii)　Using the Fourier transform determine and sketch the corresponding Fourier spectrum.

2.6　A signal with power spectral density

$$S_x(f) = \begin{cases} 1 - |f|/W & |f| < W \\ 0 & \text{elsewhere} \end{cases}$$

is passed through an ideal lowpass filter. The filter bandwidth B is intended to be set to $B = W$ but is inadvertantly set to $B = W/2$. Determine the resulting relative decrease in output signal power.

2.7　A signal has a uniform amplitude distribution given by

$$p(x) = \begin{cases} 1/4 & |x| < 2 \\ 0 & \text{elsewhere} \end{cases}$$

　　(i)　What is the relative frequency with which x exceeds the value 1.5?
　　(ii)　What is the relative frequency with which x goes outside the range $|x| < 1.5$?

2.8　For a periodic signal the *power spectrum* looks very like the *Fourier spectrum* but the weights of the line components for the former are equal to the square of the corresponding weights for the latter. Confirm this by considering a sinusoidal signal as follows:
　　(i)　Obtain a Fourier spectrum representation.
　　(ii)　Obtain a power spectrum by squaring the weights obtained in (i).
　　(iii)　Add up the weights to obtain the total power in the signal.
　　(iv)　Check this against a result obtained directly using a time integral.

2.9　The pulse signal of Problem 2.5 can be shown to correspond to

$$x(t) = \frac{1}{T} \operatorname{rect}\left(\frac{t}{T}\right) * \operatorname{rect}\left(\frac{t}{T}\right)$$

Use the convolution theorem to determine $X(f)$ and check this result with that obtained for Problem 2.5.

2.10　Use the formalism for the Fourier transform of a periodic signal together with the convolution theorem to determine the Fourier transform for the signal of Problem 2.2.

2.11 Consider the signal of Problem 2.6 corrupted by white noise with double-sided spectral density $\eta/2$. Determine the output SNR if the filter bandwidth is (i) W, (ii) $W/2$.

2.12 Consider a binary digital signal $x(t)$ given by

$$x(t) = \sum_n a_n\, p(t - nT) \equiv p(t) * \sum_n a_n \delta(t - nT)$$

where $a_n = \pm 1$ represents the data and $p(t)$ is the signalling element pulse waveform. Use the convolution theorem to show that if $P(f)$ is strictly band-limited to the interval $|f| < 1/T$ then so is $X(f)$.

3 Sinusoidal Carrier Modulation

Objectives ☐ To describe how modulation enables a message to be matched to a band pass channel.
☐ To determine the spectral occupancy of various amplitude modulated signals (DSB–SC, SSB–SC, conventional *envelope* modulation) in terms of the carrier and message spectrum.
☐ To explain how a message is recoverable from an AM signal using coherent detection.
☐ To assess the influence of phase errors on coherent detection.
☐ To show that angle (frequency or phase) modulation can be used to convey a message over a band pass channel.
☐ To assess the performance in the presence of noise of the various modulation schemes considered.
☐ To determine the spectral occupancy of a composite signal produced by frequency division multiplexing.

Introduction

This is just one example. For some purposes radio systems use a carrier frequency as low as 15 kHz while at the other extreme optical communication systems have carrier frequencies of the order of 10^{14} Hz.

The reader may be aware that radio systems generally use rather high frequencies. For example, the so-called 'citizen's band' is in the vicinity of 27 MHz. On the other hand, it has already been seen that speech signals are concentrated at relatively low frequencies, $|f| < 20$ kHz. How then can speech signals be conveyed over a radio channel? The answer is that a process called *modulation* is used to *match* the speech to the available communication channel.

This process of modulation involves the variation of some parameter of one signal (the *carrier*) by another (the *message*), the result being a modulated carrier. The modulated carrier contains complete information about the message signal and the original message can be recovered by suitable signal processing.

There are various forms of modulation. Consider a sinusoidal signal as the carrier: an information-bearing message signal can be impressed on to this by varying the amplitude, the frequency or the phase. All of these schemes are used, both separately and in combination. Only the basic methods are examined here, paying particular attention to the simplest, namely amplitude modulation.

Amplitude Modulation

There are various forms of amplitude modulation; only three are considered here: (i) double sideband suppressed carrier modulation, (ii) single sideband modulation, and (iii) conventional (envelope) amplitude modulation.

Double Sideband Suppressed Carrier Modulation

Let $m(t) = A_m\cos\omega_m t$ represent a *message* signal and $x_c(t) = A_c\cos\omega_c t$ a *carrier*, with

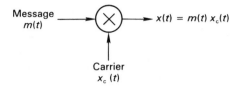

Fig. 3.1   Double sideband suppressed carrier modulation.

A multiplication process is at the heart of all amplitude modulation systems.

$F_c \gg F_m$. If the two signals are multiplied together as shown in Fig. 3.1, the following is obtained:

$$x(t) = m(t)x_c(t) = A_m\cos\omega_m t A_c\cos\omega_c t$$
$$= \frac{A_m A_c}{2} [\cos(\omega_c + \omega_m)t + \cos(\omega_c - \omega_m)t] \qquad (3.1)$$

This signal is concentrated in the vicinity of F_c and is composed of two terms known as sidebands. The term at $(F_c + F_m)$ is the *upper sideband* and the term at $(F_c - F_m)$ is the *lower sideband*. Notice that there is no component at the carrier frequency itself, hence the name *double sideband-suppressed carrier* (DSB–SC) modulation. The message information is carried in the sidebands. This modulation process can be interpreted in the frequency domain by writing the cosine waves in terms of phasors:

Recall that this representation in terms of 'positive' and 'negative' frequency components is largely a matter of mathematical convenience, a consequence of expansion in terms of complex exponentials.

$$x(t) = \frac{A_m A_c}{4} \left\{ \begin{array}{l} \exp[j(\omega_c + \omega_m)t] + \exp[-j(\omega_c + \omega_m)t] \\ + \exp[j(\omega_c - \omega_m)t] + \exp[-j(\omega_c - \omega_m)t] \end{array} \right\} \qquad (3.2)$$

This corresponds to a two-sided spectrum with positive frequency terms at $(\omega_c + \omega_m)$ and $(\omega_c - \omega_m)$ together with negative frequency terms at $-(\omega_c + \omega_m)$ and $-(\omega_c - \omega_m)$, as illustrated in Fig. 3.2.

With reference to these diagrams, note that the positive frequency part of the modulated signal is of the same form as the message spectrum but is centred on

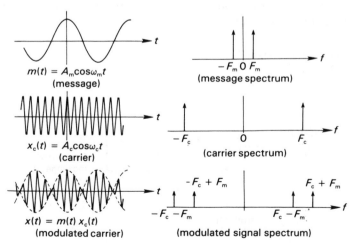

Note that we can readily interpret this in terms of the convolution theorem: multiplication in the time domain gives rise to convolution in the frequency domain.

Fig. 3.2   Time (left) and frequency (right) domain views of DSB–SC modulation.

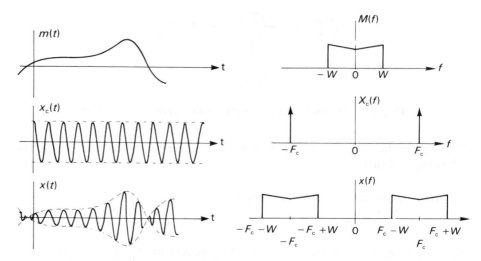

Fig. 3.3 DSB–SC modulation for a general message signal.

Using the convolution theorem:

$m(t)A_c\cos(\omega_c t)$
$\Leftrightarrow M(f) * A_c/2\,\delta(f \pm F_c)$
whence $X(f) = A_c/2\,M(f \pm F_c)$.

$f = +F_c$ rather than $f = 0$ and is scaled in amplitude by $A_c/2$. Similarly, the negative term is centred on $f = -F_c$. In essence, this modulation process has produced copies of the message spectrum at $f = \pm F_c$, with an amplitude scaling factor $A_c/2$. This is just what is required to *match* the message to a bandpass channel centred on $|f| = F_c$. The situation applies equally well to a more general message wave. Consider $m(t)$ bandlimited to $|f| < W$ modulated on to a carrier $x_c(t) = A_c\cos\omega_c t$, as shown in Fig. 3.3. Note that the modulated signal is of bandpass form with bandwidth $2W$.

This technique allows a message to be translated to a different region of frequency space for transmission over a bandpass channel. At the receiver it is necessary to be able to recover the message from the modulated signal; this can be achieved by a second multiplication process in combination with lowpass filtering. This message recovery process is called *demodulation*; a suitable receiver signal processing scheme is illustrated schematically in Fig. 3.4.

To appreciate how this demodulation process operates, consider once more a DSB–SC signal for a cosine message:

$$x(t) = A_m\cos\omega_m t A_c\cos\omega_c t$$
$$x_c(t) = A_c\cos\omega_c t$$

A multiplication process facilitates recovery of the message from an amplitude modulated signal.

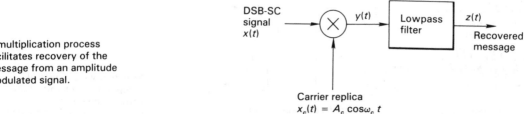

Fig. 3.4 Demodulation of a DSB–SC signal.

60

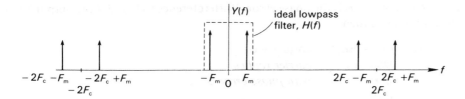

Fig. 3.5 Demodulated DSB–SC signal.

The output from the multiplier, $y(t)$, is thus

$$
\begin{aligned}
y(t) &= x(t)x_c(t) = (A_m\cos\omega_m t\, A_c\cos\omega_c t)A_c\cos\omega_c t \\
&= A_m A_c^2 \cos\omega_m t (\cos\omega_c t)^2 \\
&= A_m\cos\omega_m t\frac{A_c^2}{2}[1 + \cos(2\omega_c t)] \\
&= \frac{A_c^2}{2}A_m\cos\omega_m t + A_m\cos\omega_m t\frac{A_c^2}{2}\cos(2\omega_c t)
\end{aligned}
\tag{3.3}
$$

The first term corresponds with the message $m(t) = A_m\cos\omega_m t$ scaled in amplitude by $A_c^2/2$ and the second to a DBS–SC signal centred on $f = 2F_c$. Hence the spectrum of $y(t)$, denoted as $Y(f)$, has the form shown in Fig. 3.5.

The output spectrum is the product of $Y(f)$ and the filter transfer function $H(f)$. With $H(f)$ corresponding to an 'ideal' lowpass filter only the (scaled) message signal appears at the output:

Once again the general case is readily dealt with using the convolution theorem.

$$
z(t) = \frac{A_c^2}{2}A_m\cos\omega_m t
\tag{3.4}
$$

Once again, this result holds also for a general message spectrum, as illustrated in Fig. 3.6.

An audio signal occupies the frequency range $30\ \text{Hz} < |f| < 15\ \text{kHz}$. This signal is to be conveyed using DSB–SC amplitude modulation over a bandpass channel which transmits signals in the range 90 kHz to 120 kHz. Determine a suitable carrier frequency to match the audio signal to this channel.

Exercise 3.1

Influence of carrier phase. The availability at the receiver of a replica of the carrier to facilitate demodulation has been assumed. In practice this must be derived from the incoming DSB–SC signal — not a trivial task since the carrier is 'suppressed'. In fact, deriving a signal of the right frequency is not too difficult but this is not sufficient; it must also have the correct phase as shall be seen.

Consider the DSB–SC signal $m(t)\cos\omega_c t$ and a local carrier given by

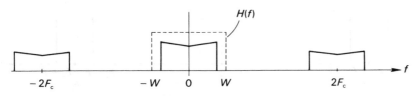

Fig. 3.6 Demodulation of a general DSB–SC signal.

$\cos(\omega_c t + \phi)$ where ϕ is the phase error. With reference to Fig. 3.4 the output from the multiplier is then

$$y(t) = [m(t)\cos\omega_c t] \quad \cos(\omega_c t + \phi)$$

$$\qquad DSB\text{-}SC \qquad carrier\ replica$$

$$\qquad signal \qquad\quad with\ phase\ error\ \phi$$

$$= \frac{m(t)}{2}[\cos\phi + \cos(2\omega_c t + \phi)]$$

$$= \frac{m(t)}{2}\cos\phi + \frac{m(t)}{2}\cos(2\omega_c t + \phi) \qquad (3.5)$$

The first term is a message scaled in amplitude by a factor of $(\cos\phi)/2$ while the second term corresponds to a DSB-SC signal in the vicinity of $2F_c$. The special case of $\phi = 0$ (no phase error) has already been considered; now consider $\phi = \pi/2$. In this case $\cos\phi = \cos\pi/2 = 0$ and there is no component in the output corresponding to the message; we simply have

$$y(t) = \frac{m(t)}{2} \cos(2\omega_c t + \pi/2) \qquad (3.6)$$

That is, a DSB-SC signal in the vicinity of $2F_c$. We conclude that the phase of the carrier reinserted at the receiver is of crucial importance. All this suggests that a rather complex receiver structure may be required if DSB-SC modulation is employed. It is appropriate, therefore, to consider other possible modulation formats which may offer the possibility of simpler receiver structures, or perhaps other advantages.

Exercise 3.2 Coherent detection of a DSB-SC signal is performed with a local oscillator, the phase of which varies by $\pm 10°$ about the optimum value. Deduce the resultant variation in the amplitude of the demodulated message.

Single Sideband Modulation

The modulation process described previously provides two complete copies of the message spectrum, one at $+F_c$ and one at $-F_c$, and both the positive and negative frequency parts of the message spectrum are preserved in each replication. As we shall see, this is redundant. It suffices to translate the positive part to the vicinity of $+F_c$ and the negative part to $-F_c$. We can establish this by considering the simple message signal

$$m(t) = A_m\cos\omega_m t$$

and a cosinusoidal carrier

$$x_c(t) = A_c\cos\omega_c t$$

DSB-SC modulation gives a spectrum as shown in Fig. 3.7.

Fig. 3.7   A DSB–SC signal.

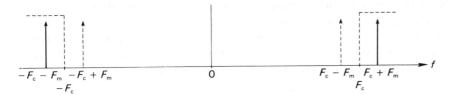

Fig. 3.8 A single sideband signal obtained by filtering a DSB–SC signal.

If this signal is passed through a highpass filter which rejects all components $|f| < F_c$ and passes all components $|f| > F_c$ a signal is obtained with spectrum as shown in Fig. 3.8.

This is the simplest and perhaps the most widely used of three basic methods for generating SSB–SC signals.

This signal corresponds to two contra-rotating phasors at $f = \pm (F_c + F_m)$ and may be expressed analytically as

$$y(t) = \frac{A_c A_m}{2}\left\{\underset{\substack{\text{negative}\\\text{frequency}\\\text{part}}}{\exp[-j(\omega_c + \omega_m)t]} + \underset{\substack{\text{positive}\\\text{frequency}\\\text{part}}}{\exp[j(\omega_c + \omega_m)t]}\right\} \qquad (3.7)$$

Considering now a more general message signal, we have the situation in Fig. 3.9. The signal is of bandlimited form with bandwidth W, one-half the bandwidth required for DSB–SC modulation. The SSB signal format is said to be *spectrally efficient*.

A more general analytic treatment of single sideband modulation makes use of the Hilbert transform which we shall not discuss here. For a particularly clear treatment see: Cattermole, K.W., *Mathematical Foundations of Communication Engineering: Vol. 1; Determinate Theory of Signals and Waves*, Pentech Press, 1985.

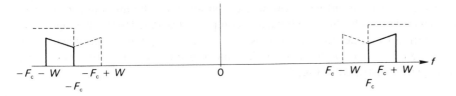

Fig. 3.9 Single sideband modulation, general message spectrum.

It must now be shown that the message is recoverable from this SSB signal. To do this consider the cosine modulation case, for which the SSB signal corresponds to a cosine wave with a frequency of $(F_c + F_m)$. We can use the same demodulation process as for DSB–SC, which involves multiplication by a replica of the carrier followed by lowpass filtering; thus

$$\text{SSB signal} \quad x(t) = \frac{A_c A_m}{2}\cos(\omega_c + \omega_m)t$$

$$\text{Carrier replica} \quad x_c(t) = A_c\cos\omega_c t$$

$$y(t) = x(t)x_c(t) = \frac{A_c^2 A_m}{2}\cos(\omega_c + \omega_m)t\cos\omega_c t$$

$$= \frac{A_c^2 A_m}{4}\left\{\cos\omega_m t + \cos(2\omega_c + \omega_m)t\right\} \qquad (3.8)$$

Again the first term corresponds to the message, which can be separated from the term at $(2\omega_c + \omega_m)$ by lowpass filtering.

Note that SSB is rather more tolerant with respect to phase errors in the reinserted carrier than is DSB–SC. Here a carrier error ϕ simply results in a similar phase shift in the demodulated message component — the amplitude remains unchanged. Some message signals, including speech, are unimpaired by such phase shifts; this can allow a significant reduction in receiver complexity.

Worked Example 3.1 Assuming a sinusoidal message, determine the influence of a phase error ϕ on demodulation of an SSB–SC signal.

Solution: The carrier *replica* is now $x_c(t) = A_c\cos(2\pi F_c t + \phi)$; hence

$$y(t) = x(t)x_c(t) = \frac{A_c^2 A_m}{4}\left[\cos(\omega_m t - \phi) + \cos(2\omega_c t + \omega_m t + \phi) \right]$$

and lowpass filtering gives

$$z(t) = \frac{A_c^2 A_m}{4} \cos(\omega_m t - \phi) \tag{3.9}$$

The recovered message is thus phase shifted by ϕ.

Conventional (Envelope) Amplitude Modulation

The two previous modulation schemes require that a replica of the carrier be available at the receiver to enable the message signal to be recovered. This complication is avoided in conventional amplitude modulation (AM) systems by making the carrier envelope the information-bearing parameter. The modulation process is again based on multiplication but a d.c. component is added to the message prior to modulation to render the signal (message + d.c. component) non-negative. The envelope of the modulated carrier then has the same shape as the message; the principle is illustrated in Fig. 3.10.

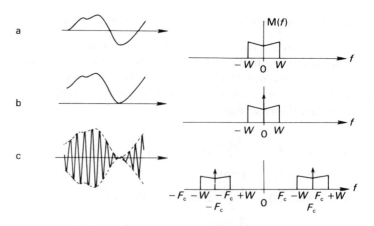

Fig. 3.10 Conventional amplitude modulation. (a) Message $m(t)$, $|m(t)| < 1$; (b) message + d.c. term, $1 + m(t)$; (c) envelope modulated signal, $x(t) = [1 + m(t)]A_c\cos\omega_c t$.

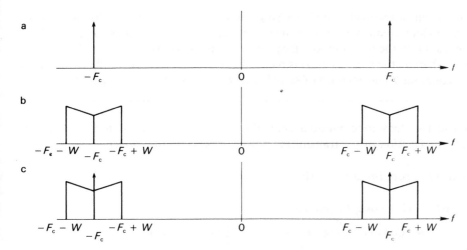

Fig. 3.11 Conventional amplitude modulation, spectral representation. (a) Unmodulated carrier; (b) DSB–SC component; (c) envelope modulated signal spectrum.

To examine this process in the frequency domain, consider $m(t) = A_m \cos\omega_m t$ with $|A_m| < 1$:

$$x(t) = [1 + A_m \cos\omega_m t]A_c \cos\omega_c t$$
$$= A_c \cos\omega_c t + A_m \cos\omega_m t A_c \cos\omega_c t \qquad (3.10)$$

The first term is simply an unmodulated carrier while the second corresponds to a DSB–SC modulated signal. The spectrum is thus the sum of the spectra for each of these signals. This result applies also for a general message as illustrated in Fig. 3.11. The modulated signal spectrum is of bandpass form with bandwidth $2W$.

The unmodulated carrier being independent of the message conveys no information. It thus constitutes 'wasted' power, so that conventional amplitude modulation makes rather inefficient use of the power available at the transmitter. On the other hand, it does provide for very simple (and hence inexpensive) receivers; it is amenable to envelope detection for which no replica of the carrier is required.

The simplicity of the receiver for *envelope* AM is the feature which resulted in its universal adoption for commercial AM broadcasting.

Envelope detection. A simple envelope detector is shown in Fig. 3.12. Assuming the diode to be ideal only the positive halves of the modulated carrier appear at the input to the lowpass filter. This signal is non-negative: it contains a d.c.

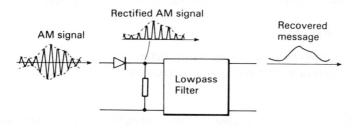

Fig. 3.12 Envelope detection.

65

component, a relatively slowly varying component corresponding to the message $m(t)$ and high frequency terms in the vicinity of the carrier frequency F_c and its harmonics. Only the d.c. and low frequency components are passed by the lowpass filter whereby the message $m(t)$ is recovered. (The d.c. term is readily removed by a.c. coupling.) This method of signal recovery is known as envelope detection.

Worked Example 3.2   Considering 'envelope' modulation and a sinusoidal message with $A_m = 1$, determine the fraction of the transmitted power 'wasted' in the carrier.

Solution: From Equation 3.10

$$x(t) = (1 + \cos\omega_m t)A_c\cos\omega_c t$$

$$= A_c\cos\omega_c t + \frac{A_c}{2}\cos(\omega_c + \omega_m)t + \frac{A_c}{2}\cos(\omega_c - \omega_m)t$$

For a sum of sine waves with different frequencies the total power is the sum of the powers in each term, giving

$$\overline{x^2(t)} = \underset{\substack{carrier \\ power}}{\frac{A_c^2}{2}} + \underset{\substack{sideband \\ power}}{\frac{A_c^2}{4}} \tag{3.11}$$

Hence 2/3 of the transmitted power is 'wasted'.

Angle Modulation

As an alternative to varying the amplitude of a carrier, its frequency or phase can be varied; both of these come under the general heading of angle modulation. This can have advantages in certain circumstances. For example, an angle modulated signal has a constant envelope which can result in more efficient use being made of the peak power capability of the transmitter. A detailed discussion is outside the scope of this text but frequency modulation (FM) is examined briefly here in view of its widespread use.

Frequency Modulation

In frequency modulation the carrier frequency is made to vary instantaneously in sympathy with the message. The instantaneous frequency f_i is given by

$$f_i = F_c + K_f m(t) \tag{3.12}$$

where K_f is a modulation constant which determines the peak deviation of the carrier from its centre frequency F_c. A frequency modulated wave is illustrated in Fig. 3.13.

The FM signal is not readily amenable to spectral analysis but the bandwidth of the modulated signal can be estimated as follows: if F_d is the peak frequency deviation

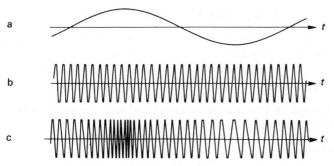

Fig. 3.13   Frequency modulation process. (a) Message; (b) unmodulated carrier; (c) frequency modulated signal.

then the instantaneous frequency varies between $F_c - F_d$ and $F_c + F_d$. This suggests that a transmission bandwidth B of the order of $2F_d$ may be required for a channel to pass the FM signal. But this is independent of the bandwidth of the message — an unlikely result. In fact, it transpires that the bandwidth is not independent of the message but is given approximately by

$$B = 2F_d + 2W \tag{3.13}$$

This is known as *Carson's Rule* for FM bandwidth.

Many systems employ $F_d \gg W$ and the FM bandwidth requirement can exceed considerably that for AM. Hence FM is not *spectrally efficient*. However, if the required bandwidth is available, FM has the capability to provide a very high quality transmission system (e.g. FM broadcasting).

FM Signal Analysis

An FM signal is described analytically by

$$x(t) = \cos \Phi = \cos \left(2\pi F_c t + 2\pi K_f \int_0^t m(u) \, du \right) \tag{3.14}$$

$$f_i = \frac{1}{2\pi} \frac{d\Phi}{dt} = F_c + K_f m(t) \tag{3.15}$$

It is the ability of FM to provide very high quality transmission which has resulted in its widespread use for high-fidelity (Hi-Fi) and stereophonic broadcasting. The spectral inefficiency dictates the use of very high carrier frequencies (e.g. VHF in the region of 100 MHz in the UK).

in accord with Equation 3.12. Note that with $m(t)$ normalized to have a peak value of unity, $|m(t)| < 1$, K_f corresponds to the peak instantaneous frequency deviation, F_d.

Considering the special case of $m(t) = \cos \omega_m t$ we have

$$x(t) = \cos \left(\omega_c t + 2\pi F_d \int_0^t \cos(\omega_m u) \, du \right)$$

$$= \cos \left(\omega_c t + \frac{2\pi F_d}{\omega_m} \sin \omega_m t \right) = \cos(\omega_c t + \beta \sin \omega_m t) \tag{3.16}$$

where $\beta = F_d / F_m$ is the deviation ratio, the ratio of the peak instantaneous frequency deviation to the frequency of the modulating signal. Using complex

For the general case of an arbitrary lowpass message signal with bandwidth W the deviation ratio is defined by F_d / W.

exponentials we can express the FM signal as follows:

$$x(t) = \tfrac{1}{2}\exp(j\omega_c t)\exp(j\beta \sin \omega_m t) + \tfrac{1}{2}\exp(-j\omega_c t)\exp(-j\beta \sin \omega_m t) \qquad (3.17)$$

terms associated   terms associated
with + ve frequency   with − ve frequency
carrier component   carrier component

Concentrating on just the terms associated with the positive frequency carrier component we note that

$$\exp(j\beta \sin \omega_m t)$$

is periodic and may thus in principle be represented by a Fourier series

$$\exp(j\beta \sin \omega_m t) = \sum_n c_n \exp(jn\omega_m t) \qquad (3.18)$$

where the Fourier coefficients c_n depend on β. It follows that

$$\tfrac{1}{2}\exp(j\omega_c t)\exp(j\beta \sin \omega_m t) = \tfrac{1}{2}\exp(j\omega_c t)\sum_n c_{n|}\exp(jn\omega_m t) \qquad (3.19)$$

defines a set of spectral lines centred on $f = F_c$, with line spacing F_m and relative strengths depending on β. By addressing the terms in Equation 3.17 associated with the negative frequency carrier component a similar set of spectral lines may be identified centred on $f = -F_c$. Taken together these two sets of lines provide a frequency domain description for the FM signal with a cosine modulating signal. Notice that it is much more complicated than for the various forms of amplitude modulation considered earlier in this chapter. Even for this simple case there are in principle an infinite number of spectral components to contend with. How many of these need in practice be considered depends on the relationship between c_n and β. This is given by

Bessel functions have their origin in the studies of planetary motion conducted by F.W. Bessel (1784–1846).

$$c_n = J_n(\beta)$$

where $J_n(\cdot)$ represents a set of special functions, known as Bessel functions. It is found that $|c_n|$ reduces quite rapidly for $|n| > \beta$ and taking as significant spectral lines of order $-(1 + \beta) \leq n \leq (1 + \beta)$ is generally appropriate. Since the lines are spaced in frequency by F_m this indicates a transmission bandwidth requirement of $\pm(1 + \beta)F_m$ around the carrier frequency. This suggests

$$B = 2F_m(1 + \beta) = 2F_m + 2F_d \qquad (3.20)$$

which is consistent with Equation 3.13 if we associate F_m with the message bandwidth W.

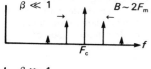
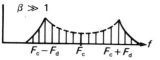

Illustrative FM spectra.

Exercise 3.3 The FM broadcasting system adopted in the United Kingdom employs a frequency deviation of ± 75 kHz. The message bandwidth is 15 kHz for monophonic transmission and 53 kHz for stereo. Determine the approximate bandpass channel bandwidth required in each case. Compare this with the bandwidth requirements for DSB–SC.

Considering now demodulation, a particularly simple scheme makes use of a frequency selective network such that the frequency variations give rise also to amplitude variations. The message can then be recovered by using an envelope detector. This scheme is illustrated schematically in Fig. 3.14 and may be appre-

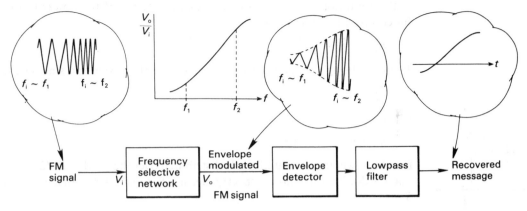

Fig. 3.14 Slope detection of FM signal.

ciated as follows: when the input signal instantaneous frequency is low, say $f = f_1$, the output from the frequency selective network is small. On the other hand, if the input instantaneous frequency is high, $f = f_2$, the output is large. As the instantaneous frequency varies between these extremes, in sympathy with the message, so the amplitude of the output varies giving rise to 'envelope' modulation. The message may now be recovered using an envelope detector. This detection scheme for frequency modulation is known as *slope detection*. Other methods of FM detection are available; however, the above should suffice to allow the principle of frequency modulation and demodulation to be appreciated.

The most widely adopted demodulation scheme makes use of a phase-locked loop, a circuit of widespread application in communication systems. It is discussed briefly in Chapter 4 and more fully in Gardner, F.M., *Phase Lock Techniques*, Wiley, 1966.

Communication in the Presence of Noise

We shall now examine and compare the performance in the presence of noise of various modulation schemes. To this end we consider first the representation of bandpass filtered noise.

Bandpass Noise

Consider white noise with double-sided spectral density $\eta/2$ applied to a bandpass filter centred on F_0 with bandwidth B_T. The resultant bandpass noise has total power ηB_T concentrated in the vicinity of F_0.

An appropriate *time domain* model for bandpass noise is a sinewave with nominal frequency F_0 with randomly varying amplitude and phase

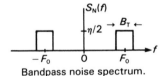

Bandpass noise spectrum.

$$n(t) = r_n(t)\cos(\omega_0 t + \phi_n(t)) \tag{3.21}$$
$$= r_n(t)\{\cos\omega_0 t \cos\phi_n(t) - \sin\omega_0 t \sin\phi_n(t)\}$$
$$= n_1(t)\cos\omega_0 t + n_2(t)\sin\omega_0 t \tag{3.22}$$

where

$$n_1(t) = r_n(t)\cos\phi_n(t) \tag{3.23a}$$
$$n_2(t) = -r_n(t)\sin\phi_n(t) \tag{3.23b}$$

Here $n_1(t)$ and $n_2(t)$ are lowpass functions occupying the frequency range $|f| \leq B_T/2$ such that when effectively DSB-SC modulated on to the cosine and sine 'carriers' as in Equation 3.22 the resultant bandpass noise falls in the band $F_0 - B_T/2 \leq |f| \leq F_0 + B_T/2$.

The total noise power can be expressed in terms of the power in the lowpass functions, $n_1(t)$, $n_2(t)$ as follows:

$$\overline{n^2} = \tfrac{1}{2}\,\overline{n_1^2} + \tfrac{1}{2}\,\overline{n_2^2}$$

There is, though, equal power associated with the cosine and sine terms, so that

$$\overline{n_1^2} = \overline{n_2^2}$$

whence

$$\overline{n^2} = \overline{n_1^2} = \overline{n_2^2} = \eta B_{\mathrm{T}} \tag{3.24}$$

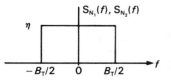

Equivalent lowpass noise power spectral density is η W/Hz.

But the power in $n_1(t)$ is uniformly distributed over the band $-B_{\mathrm{T}}/2 < f < B_{\mathrm{T}}/2$ so that the double-sided power spectral density $S_{\mathrm{N}_1}(f)$ has strength η, as similarly does $S_{\mathrm{N}_2}(f)$;

$$S_{\mathrm{N}_1}(f) = S_{\mathrm{N}_2}(f) = \eta, \ |f| < B_{\mathrm{T}}/2 \tag{3.25}$$

Baseband Transmission

Consider baseband transmission for a message $m(t)$ having bandwidth W, mean power $P_{\mathrm{S}} = \overline{m^2}$ and normalized to $|m(t)| \leq 1$. If this is corrupted by white noise with double-sided spectral density $\eta/2$ the signal plus noise waveform may be lowpass filtered by an ideal lowpass filter with bandwidth W. This filter has no effect on the signal but rejects out of band noise components so that the noise power at the filter output is just ηW. The signal to noise ratio is thus

$$\mathrm{SNR}_{\mathrm{O}_{\text{baseband}}} = \frac{\overline{m^2}}{\eta W} = \frac{P_{\mathrm{S}}}{\eta W} = \gamma \tag{3.26}$$

We shall take this as a reference against which to assess the performance of modulated systems.

DSB-SC Transmission

The transmitted signal has the form

$$x(t) = m(t)\cos \omega_c t \tag{3.27}$$

with average power $P_{\mathrm{S}} = \overline{m^2}/2$ where for convenience we have normalized the carrier amplitude to unity. At the output of the bandpass filter we have signal plus noise, with the noise as represented in Equation 3.22 having average power $\eta B_{\mathrm{T}} = 2\eta W$ since the transmission bandwidth required for DSB-SC modulation is $2W$. At this point the signal to noise ratio prior to demodulation, SNR_i, is given by

$$\mathrm{SNR}_{\mathrm{i}_{\text{DSB-SC}}} = \frac{P_{\mathrm{S}}}{2\eta W} = \frac{\gamma}{2} \tag{3.28}$$

Demodulating as shown in Fig. 3.4 using a carrier replica $\cos \omega_c t$ we obtain at the output of the multiplier

$$\begin{aligned}
y(t) &= \{m(t)\cos \omega_c t + n(t)\}\cos \omega_c t \\
&= \{m(t)\cos \omega_c t + n_1(t)\cos \omega_c t + n_2(t)\sin \omega_c t\}\cos \omega_c t \\
&= \{m(t) + n_1(t)\}\cos^2 \omega_c t + n_2(t)\sin \omega_c t \cos \omega_c t \\
&= \frac{m(t) + n_1(t)}{2}(1 + \cos 2\omega_c t) + \frac{n_2(t)}{2}\sin 2\omega_c t
\end{aligned} \tag{3.29}$$

Terms centred on $2\omega_c$ are removed by the lowpass filter giving

$$z(t) = \tfrac{1}{2}m(t) + \tfrac{1}{2}n_1(t)$$

so that the output signal to noise ratio is

$$\text{SNR}_{\text{o}_{\text{DSB-SC}}} = \frac{\tfrac{1}{4}\overline{m^2}}{\tfrac{1}{4}\overline{n_1^2}} = \frac{\overline{m^2}}{\overline{n_1^2}} \tag{3.30}$$

But from Equation 3.25 we have $\overline{n_1^2} = \overline{n^2} = \eta B_T = 2\eta W$ while the received modulated signal power is $P_S = \overline{m^2}/2$. We thus have

$$\text{SNR}_{\text{o}_{\text{DSB-SC}}} = \frac{\overline{m^2}}{2\eta W} = \frac{P_S}{\eta W} = \gamma \tag{3.31}$$

We note that while prior to demodulation the SNR was $\gamma/2$, 3 dB less than for baseband transmission, the SNR after demodulation is γ, the same as for baseband transmission. This improvement in SNR between the input and output of the demodulator can be associated with the rejection of the quadrature noise component $n_2(t)$ which carries one-half of the bandpass noise power.

An alternative, equivalent, interpretation makes use of the observation that the upper and lower sidebands of the DSB-SC signal are directly related to one another — corresponding to the positive and negative frequency components of the message — while the noise components either side of F_c are unrelated. Consequently on demodulation the positive and negative frequency signal voltage components add up while the unrelated noise terms combine on an 'addition of power' basis. This voltage addition of signal terms but power addition of noise terms yields the noted 3 dB improvement in SNR.

SSB Transmission

Considering upper sideband modulation as shown in Fig. 3.9 we note that the signal and noise are concentrated in the band F_c to $F_c + W$. The signal power is P_S and the noise power is ηW whence

$$\text{SNR}_{\text{i}_{\text{SSB}}} = \frac{P_S}{\eta W} = \gamma \tag{3.32}$$

Demodulation has the effect of translating the positive frequency sideband into the frequency range $0 \leq f \leq W$ and the negative frequency sideband into the range $-W \leq f \leq 0$. There is thus no overlap of spectral components and the question of voltage or power addition does not arise: signal and noise are processed identically so that the output SNR_o is the same as the input SNR_i prior to demodulation

$$\text{SNR}_{\text{o}_{\text{SSB-SC}}} = \text{SNR}_{\text{i}_{\text{SSB-SC}}} = \frac{P_S}{\eta W} = \gamma \tag{3.33}$$

Conventional (Envelope) AM Transmission

Here, as for DSB-SC, the modulated signal occupies a bandwidth $B_T = 2W$ so that the bandpass noise power is $2\eta W$ and with total signal power P_S we have, prior to demodulation,

$$\text{SNR}_{\text{i}_{\text{Envelope AM}}} = \frac{P_S}{2\eta W} = \frac{\gamma}{2} \tag{3.34}$$

See Worked Example 3.2.

The signal *sidebands* and the noise are processed essentially as for DSB-SC. However, only a certain fraction of the total transmitted power is concentrated in the information-bearing sidebands, the remainder being carrier power. Hence, if k is the ratio of sideband power to total power we have

$$\mathrm{SNR}_{\mathrm{O_{Envelope\,AM}}} = k\gamma \tag{3.35}$$

If envelope detection is employed then this argument holds provided $\mathrm{SNR}_i \gtrsim 10$ dB.

Typically $k \ll 1$ so that considerably more signal power is required to achieve a given SNR_o with envelope AM than with DSB-SC or SSB-SC modulation.

Frequency Modulation

Consider the FM signal of Equation 3.14

$$x(t) = \cos\left(2\pi F_c t + 2\pi K_f \int_0^t m(u)\,\mathrm{d}u\right) \tag{3.36}$$

occupying a bandwidth $B_T = 2F_d + 2W$ and with signal power $P_S = \frac{1}{2}$. For this case of a general message signal $m(t)$ we define the deviation ratio β as the ratio of the peak instantaneous frequency deviation F_d to the maximum message frequency W. Since the message is normalized to unit peak amplitude, $|m(t)| \leq 1$, then from Equation 3.15 F_d corresponds to K_f and we have

$$\beta = \frac{F_d}{W} = \frac{K_f}{W} \tag{3.37}$$

and

Carson's rule for FM bandwidth.

$$B_T = 2F_d + 2W = 2W(1 + \beta) \tag{3.38}$$

The input noise power is thus $2\eta W(1 + \beta)$ so that the signal to noise ratio prior to demodulation — often referred to as the carrier to noise ratio CNR — is given by

$$\mathrm{SNR}_{i_{FM}} = \mathrm{CNR} = \frac{P_S}{2\eta W(1 + \beta)} = \frac{\gamma}{2(1 + \beta)} \tag{3.39}$$

This is necessarily less than γ and markedly so for $\beta \gg 1$.

Considering now the demodulation process for the FM signal in the absence of noise the receiver must effectively perform differentiation of the phase deviation to recover a measure of the instantaneous frequency deviation produced by the message. Denoting as $y(t)$ the recovered signal we have

$$\begin{aligned}
y(t) &= \frac{1}{2\pi} \frac{\mathrm{d}}{\mathrm{d}t}\left\{2\pi K_f \int_0^t m(u)\,\mathrm{d}u\right\} \\
&= \frac{1}{2\pi}\{2\pi K_f m(t)\} = K_f\, m(t)
\end{aligned} \tag{3.40}$$

so that the output signal power S_o is

$$S_o = K_f^2 \overline{m^2} \tag{3.41}$$

Turning our attention now to the noise we shall for convenience consider an unmodulated carrier perturbed by bandpass noise

$$\cos \omega_c t + n(t) = \cos \omega_c t + n_1(t) \cos \omega_c t + n_2(t) \sin \omega_c t$$
$$= (1 + n_1(t)) \cos \omega_c t + n_2(t) \sin \omega_c t \qquad (3.42)$$

This composite carrier-plus-noise waveform has instantaneous phase perturbation $\phi(t)$ given by

$$\tan \phi(t) = \frac{n_2(t)}{1 + n_1(t)} \qquad (3.43)$$

Assuming now that the noise is much smaller than the carrier we can reasonably make use of the approximation $n_1(t) \ll 1$ to obtain

$$\tan \phi(t) \simeq n_2(t) \qquad (3.44)$$

and noting that similarly $n_2(t) \ll 1$ we have

$$\tan \phi(t) \simeq \phi(t) \simeq n_2(t) \qquad (3.45)$$

The instantaneous frequency associated with this noise-induced phase perturbation is given by

$$\frac{1}{2\pi} \frac{\mathrm{d}\phi(t)}{\mathrm{d}t} = \frac{1}{2\pi} \frac{\mathrm{d}}{\mathrm{d}t} n_2(t) \qquad (3.46)$$

Now differentiation of a Fourier transformable time domain function has the effect of multiplying the transform by $j2\pi f$. Since power is proportional to voltage squared then for a finite average power signal the power spectral density is multiplied by $|j2\pi f|^2 = (2\pi)^2 f^2$. Hence given the differentiated noise process of Equation 3.46 we obtain as the output noise power spectral density function

$$S_{N_o}(f) = \left(\frac{1}{2\pi}\right)^2 \{(2\pi)^2 f^2 S_{N_2}(f)\}$$

$$= \begin{cases} f^2 \eta & |f| \le B_T/2 \\ 0 & \text{elsewhere} \end{cases} \qquad (3.47)$$

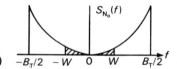

using $S_{N_2}(f)$ as given by Equation 3.25.

Only noise falling in the band $|f| \le W$ will be passed by the final lowpass filter so that the output noise power N_0 is

$$N_o = \int_{-W}^{W} S_{N_o}(f) \, \mathrm{d}f$$

$$= \int_{-W}^{W} f^2 \eta \, \mathrm{d}f = \eta \left[\frac{f^3}{3} \right]_{-W}^{W}$$

$$= \tfrac{2}{3} \eta W^3 \qquad (3.48)$$

Combining Equations 3.41 and 3.48 we obtain as the output signal to noise ratio

$$\text{SNR}_{\text{OFM}} = \frac{S_o}{N_o} = \frac{K_f^2 \overline{m^2}}{\tfrac{2}{3} \eta W^3}$$

$$= \frac{3}{2} \left(\frac{K_f}{W} \right)^2 \frac{\overline{m^2}}{\eta W}$$

$$= 3\beta^2 \overline{m^2} \frac{\tfrac{1}{2}}{\eta W}$$

$$= 3\beta^2 \overline{m^2} \gamma \qquad (3.49)$$

73

Notice that this will be greater than γ provided we select $\beta > 1/(3\overline{m^2})$. We conclude that for a given available signal power frequency modulation can provide an SNR advantage compared with DSB-SC, SSB-SC and baseband transmission which is given by

$$\frac{\text{SNR}_{\text{o}_{\text{FM}}}}{\gamma} = 3\beta^2\overline{m^2} \tag{3.50}$$

We note also that FM demodulation can provide a very significant improvement compared with the SNR prior to demodulation.

$$\frac{\text{SNR}_{\text{o}_{\text{FM}}}}{\text{SNR}_{\text{i}_{\text{FM}}}} = \frac{3\beta^2\overline{m^2}\,\gamma}{\gamma/\{2(1+\beta)\}}$$
$$= 6\beta^2(1+\beta)\overline{m^2} \tag{3.51}$$

It is tempting to conclude that by employing a sufficiently large value of β an arbitrarily large SNR_o could be obtained for a given FM signal power and noise density but a word of caution is appropriate. For the performance indicated here to be realized an adequate input CNR is required otherwise noise can 'capture' the FM demodulator. This noise capture process is related to the failure at low CNRs of the approximations indicated in Equations 3.44 and 3.45; instead of noise representing a small phase perturbation on the carrier the noise becomes dominant. This produces the FM threshold effect whereby SNR_o degrades rapidly once the CNR is reduced below some critical value in the region of 10 to 20 (10 dB to 13 dB).

Frequency Division Multiplexing

Frequently it is desired to combine several messages into a composite signal for transmission over a single communication channel. This is the case, for example, with intercontinental communication for which many telephone conversations are combined and then transmitted via a submarine cable or a communications satellite. This can be achieved with the aid of frequency division multiplexing (FDM). The various messages are amplitude modulated on to different carrier frequencies and the modulated signals are added together to form a composite message signal. Provided the carrier separation is sufficient to ensure that there is no spectral overlap of the sidebands of one signal with those of another the messages can be separated at the receiver by bandpass filtering. They can then be demodulated in the usual way. Any of the forms of amplitude modulation mentioned can be employed in principle but often power economy dictates that a supressed carrier (DSB–SC or SSB–SC) scheme be adopted.

In the interests of bandwidth economy SSB–SC is preferred. The principle is illustrated in Fig. 3.15 using a single-sided spectral representation. The channel spacing is slightly greater than the message information bandwidth W to facilitate channel separation filtering at the receiver. Once an individual SSB channel signal has been separated, demodulation in the usual way allows the message to be recovered.

Quite complex multiplexing and modulation formats are encountered in practical telecommunication systems. For example, the CCITT (Consultative Committee of the International Telecommunications Union — an international

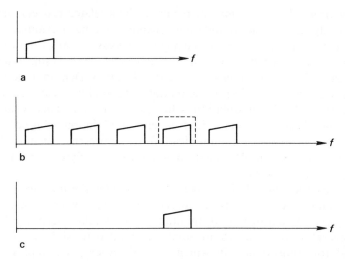

Fig. 3.15 Frequency division multiplexing/demultiplexing. (a) Representative message signal spectrum; (b) composite FDM signal spectrum; (c) separated channel, ready for demodulation.

advisory body) recommend an FDM hierarchy for telephony as shown in Table 3.1. This structure allows telephone calls to be combined in blocks of appropriate sizes for transmission through a national or international telecommunications network.

Telephone speech is concentrated in the band 300 Hz to 3·4 kHz. To allow for finite cut-off filters, however, a nominal channel allocation of 0 to 4 kHz is usually assumed, leading to a 4 kHz channel spacing in an FDM assembly.

Table 3.1   CCITT FDM Hierarchy

Channel		1 channel (4 kHz)
Group		12 channels (48 kHz)
Supergroup	5 groups	60 channels (240 kHz)
Mastergroup	5 supergroups	300 channels (1.2 MHz)
Supermastergroup	3 mastergroups	900 channels (3.6 MHz)

When considering the transmission of information through networks we often refer to *traffic* flowing through, or carried by, the network in analogy with vehicular traffic flowing along roads. One of the attributes of multiplexing is that it enables us to combine traffic (e.g. telephone calls) into suitably sized units for efficient and economic transmission.

It should be noted that an FDM signal may be subjected to further modulation processes in transmission. For example, it is not uncommon to combine several telephone channels using FDM based on SSB–SC modulation and then to use frequency modulation to transmit this composite signal over a radio or satellite link.

Summary

It has been shown that by the process of modulation a message signal can be *translated* in frequency space and thereby rendered compatible with a bandpass channel. The original message may be recovered by a related process, demodulation. Specifically, amplitude modulation schemes have been examined in some detail and we have noted that the fundamental process is that of multiplication; amplitude modulation is achieved by multiplying a message by a sinusoidal carrier and, in the case of SSB, by filtering the resultant signal. Demodulation involves similar processes; multiplication of the modulated signal by a carrier replica and lowpass filtering. Angle modulation schemes, such as frequency and phase modulation, have been mentioned with frequency modulation being considered in some detail. The problem of communication in the presence of noise has been considered by way of simple signal to noise ratio analyses for each of the modulation schemes introduced.

The purpose of all modulation schemes is to render the message compatible with the available channel. This is especially important when a large number of messages need to be transmitted simultaneously. We have seen that by modulating the several messages on to different carriers it is possible to form a composite signal for transmission in which the constituents are kept separate in frequency space. This process, known as frequency division multiplexing, is in widespread use. It is employed, for example, in long distance telephony for which messages are combined in accordance with CCITT guidelines. Also, albeit in a rather less structured way, it is essentially what is involved in broadcast radio. The various stations are allocated different 'frequencies' to allow them to be received separately.

Finally, we note that in this chapter we have assumed the use of a *sinusoidal* carrier. While this is certainly commonplace it is by no means universal. There are, as is seen later, various forms of pulse modulation which have a part to play in telecommunication systems.

Problems

3.1 Consider conventional amplitude modulation of a sinusoidal carrier $x_c(t)$ with a sinusoidal message $m(t)$:

$$x_c(t) = A_c \cos\omega_c t \qquad\qquad m(t) = A_m \cos\omega_m t$$

Determine the fraction of the total transmitted power concentrated in the modulation sidebands for (i) $A_m = A_c$; (ii) $A_m = A_c/2$; (iii) $A_m = aA_c$; $|a| < 1$.

For a speech message it is generally required that the r.m.s. value of the message must not exceed about one-tenth of the peak carrier value. Assuming that the same general behaviour noted above for sinusoidal messages holds also for the more complex speech signal, estimate the fraction of the transmitted power concentrated in the message sidebands for speech modulation at the level suggested.

3.2 By considering a sinusoidal message $A_m \cos\omega_m t$ examine the influence of a local oscillator *frequency* error at the receiver for the detection of (i) a DSB–SC signal, and (ii) an SSB–SC signal.

Hence suggest why for some signals, such as speech, SSB–SC may be subjectively more tolerant to frequency errors than is DSB–SC.

3.3 Twelve speech signals are to be combined using frequency division multiplexing for transmission over a wideband radio link. Each signal is concentrated in 300 Hz $< |f| <$ 3.4 kHz and is allocated a nominal lowpass channel bandwidth of 4 kHz to allow for the use of practical filters with finite transition regions.

Estimate the radio bandwidth required if the signals are combined using SSB–SC frequency division multiplexing and the radio receiver employs conventional (envelope) amplitude modulation.

4 Radio Receiver Principles

Objectives
- ☐ To introduce the concept of selectivity.
- ☐ To explain the principle of a tuned radio frequency (TRF) receiver.
- ☐ To explain the principle of the superheterodyne (superhet) receiver.
- ☐ To introduce and distinguish between adjacent channel and image rejection.
- ☐ To note the criteria influencing selection of the intermediate frequency for a superhet receiver.
- ☐ To note briefly the double conversion receiver principle.
- ☐ To describe the principle of automatic frequency control (AFC).

In the previous chapter it was explained how it is possible to modulate a carrier with a message and subsequently to recover the message from the modulate carrier. In a radio system, the modulation process is performed at the transmitter, and the modulated signal is then amplified and broadcast as a radio signal — a propagating electromagnetic wave field. At the receiver this signal is picked up by a receiving *aerial* or *antenna* and it is the task of the radio receiver to process this signal so that it can be applied to a demodulator for message recovery. The details of radio propagation and the physical mechanism whereby an aerial can pick up the radio signal are not dealt with here. Receiver system principles and the associated signal processing operations are considered.

It should be noted first that radio communication involves a form of frequency division multiplexing in that different radio channels in a given geographical location use separate carrier frequencies. These carriers are sufficiently well separated in frequency to ensure that, with the aid of a tuned circuit, a receiver can be adjusted (*tuned*) to pass the carrier and modulation sidebands of a wanted signal while rejecting substantially the unwanted signals.

Worked Example 4.1

A simple radio receiver employs a single tuned circuit for channel selection; the relative response is given approximately by

Note the use of $|H(f)|^2$ because we are interested in the power which is proportional to (voltage)^2.

$$|H(f)|^2 = \frac{1}{1 + (f - f_0)^2/B^2} \tag{4.1}$$

where f_0 = centre frequency and $B_{RF} = 2B \equiv -3$ dB bandwidth. If two equally strong, co-located stations transmit at f_0 and at $f_1 = f_0 + 7B$, determine the ratio at the input to the demodulator of the wanted signal to interference power, expressed in decibels.

Solution: The ratio of the wanted signal to interference power after filtering is given by

$$\frac{P_{\text{wanted}}}{P_{\text{interference}}} = \frac{|H(f_0)|^2}{|H(f_1)|^2} \tag{4.2}$$

$$= 1 + (f_1 - f_0)^2/B^2$$
$$= 1 + (7B)^2/B^2 = 50$$
$$= 10 \log_{10}(50) \equiv 17 \text{ dB}$$

One of the tasks of a radio receiver is to effect this selection of the wanted from the unwanted signals. A useful measure of the *selectivity* of a receiver is provided by the ratio of the bandwidth at different levels on the relative response curve. The -40 dB to -3 dB bandwidth ratio is often considered, for example. Note that a large ratio implies poor selectivity, as illustrated in Fig. 4.1. An ideal receiver would have a selectivity ratio of unity.

This would require an *ideal* bandpass filter.

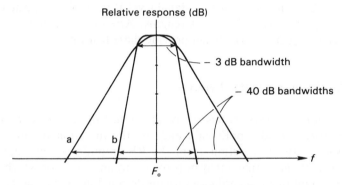

Fig. 4.1   Low (a) and high (b) selectivity receivers.

Determine the selectivity ratio for the receiver of Worked Example 4.1.

Exercise 4.1

Amplification of the radio frequency signal — RF amplification — is also usually required since the received signal can be of rather low amplitude. The amount of RF amplification required depends on the strength of the received signal which in turn depends on the distance of the receiver from the transmitter and on the effectiveness of the receiving aerial. To allow for variations in signal strength a variable gain amplifier is employed in conjunction with an automatic feedback loop. This *automatic gain control* (AGC) system acts to keep the signal at the input to the demodulator at a suitable predetermined average level.

For details of feedback system analysis see: Horrocks, D.H., *Feedback Circuits and Op Amps*, Van Nostrand Reinhold, 1983.

Consider a line-of-sight radio system in which a receiver is required to operate satisfactorily at distances from the transmitter varying from 1 km to 100 km. Estimate the AGC range required, expressing this in dB.

Worked Example 4.2

Solution: Since power density at the receiver varies with the square of the distance from the transmitter the ratio of the received powers is given by

$$\frac{P(d_{min})}{P(d_{max})} = \left(\frac{d_{max}}{d_{min}}\right)^2 \tag{4.3}$$

where d_{min} and d_{max} are the minimum and maximum transmitter-receiver distances and $P(d_{min})$, $P(d_{max})$ are the corresponding received signal powers.

Here:

$$\frac{P(1\ \text{km})}{P(100\ \text{km})} = \left(\frac{100}{1}\right)^2 = 10^4$$
$$\equiv 10\ \log_{10}(10^4) = 40\ \text{dB}$$

The AGC system must thus have a *dynamic range*, defined as the difference between the maximum and minimum gains expressed in dB, of at least 40 dB.

There are thus three main signal processing operations to be performed prior to demodulation: filtering for channel selection; amplification to ensure an adequate signal; and AGC to ensure a constant signal level at the input to the demodulator for a wide range of received signal strengths. Attention can now be turned to possible receiver structures which can provide these features.

Tuned Radio Frequency (TRF) Receiver

The simplest radio receivers are based directly on the three processes noted above. An RF amplifier provides both gain and, by the inclusion of tuned circuits, frequency selectivity. The output from the RF amplifier is then applied to a demodulator. Level adjustment may be provided by an AGC loop which varies the gain of the RF amplifier in accordance with the detected signal level at the output of the demodulator. Such a receiver is shown schematically in Fig. 4.2, where for the sake of definiteness envelope amplitude modulation has been assumed. This is a very simple receiver (so much so that the AGC loop may well be omitted and signal level adjustments effected manually) but it is not widely used. The main reason for this is that the tuned circuits in the various stages of the RF amplifier must remain properly aligned with one another as we tune the receiver over its operating range in order to select different frequency transmissions. This is generally achieved with the aid of a ganged capacitor, which is actually several variable capacitors all sharing a common control shaft. This becomes increasingly inconvenient as more stages are incorporated to improve selectivity or provide increased gain and in practice there is a limit to the number of stages which can be kept in step in this way. TRF receivers thus tend to provide rather poor selectivity.

Broadly similar tracking problems arise with electronically tuned receivers based on voltage-variable capacitance diodes if many stages are involved.

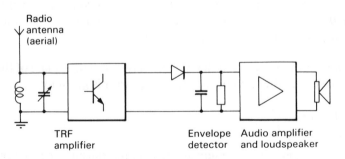

Fig. 4.2 TRF receiver.

Superheterodyne (Superhet) Receivers

Most practical radio receivers employ the superheterodyne principle in which the incoming signal centred on F_c is translated (or *converted*) in frequency to a fixed *intermediate frequency* (IF) band. The necessary frequency translation of the input signal is achieved by multiplication in the time domain by a fixed amplitude sinusoidal signal with frequency F_{LO} offset from the carrier by an amount F_{IF}. This is illustrated schematically in Fig. 4.3. The principle may be appreciated as follows:

Consider, by way of example, a DSB–SC signal:

$$x(t) = m(t)\cos(2\pi F_c t) \tag{4.4}$$

multiplied by a constant amplitude sinusoid provided by an oscillator within the receiver, known as a local oscillator:

$$x_{LO}(t) = A\cos(2\pi F_{LO}t) \tag{4.5}$$

The output from the multiplier is given by

$$
\begin{aligned}
y(t) &= m(t)\cos(2\pi F_c t)A\cos(2\pi F_{LO}t) \\
&= \frac{Am(t)}{2}\{\cos[2\pi(F_c + F_{LO})t] + \cos[2\pi(F_c - F_{LO})t\} \\
&= m(t)\frac{A}{2}\cos[2\pi(F_c + F_{LO})t] + m(t)\frac{A}{2}\cos(2\pi F_{IF}t) \\
&= \text{DSB–SC signal} \qquad\qquad \text{DSB–SC signal} \\
&\quad\ \ \text{centre on} \qquad + \qquad \text{centred on} \\
&\quad\ \ f = F_c + F_{LO} \qquad\quad f = F_{IF} = |F_c - F_{LO}|
\end{aligned}
\tag{4.6}
$$

If this signal is applied to a multi-stage bandpass amplifier with a response centred on the fixed intermediate frequency $|F_c - F_{LO}|$, the terms centred on $|F_c + F_{LO}|$ are heavily attenuated while the terms at $F_{IF} = |F_c - F_{LO}|$ are amplified. The output from the bandpass amplifier is thus

$$z(t) = m(t)\frac{A}{2}B\cos(2\pi F_{IF}t) \tag{4.7}$$

where B represents the voltage gain of the IF amplifier, and this signal can be demodulated as discussed in Chapter 3.

The frequency conversion process does not require a perfect multiplier; it is

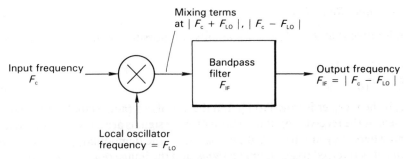

Fig. 4.3   Superheterodyne principle.

sufficient for the modulated carrier and local oscillator signals to be combined in a suitable *non-linear* manner to produce the required sum and difference frequency components. The circuit is generally referred to as a *mixer*. To illustrate this we consider a square-law device operating on $x_c(t) + x_{LO}(t)$ with output $y(t)$ given by

$$
\begin{aligned}
y(t) &= [x_c(t) + x_{LO}(t)]^2 \\
&= [m(t) \cos(2\pi F_c t) + A \cos(2\pi F_{LO} t)]^2 \\
&= m^2(t) \cos^2(2\pi F_c t) + A^2 \cos^2(2\pi F_{LO} t) + 2m(t)A \cos(2\pi F_c t) \cos(2\pi F_{LO} t) \\
&= m^2(t) \frac{1}{2}[1 + \cos(4\pi F_c t)] \\
&\quad + A^2 \frac{1}{2}[1 + \cos(4\pi F_{LO} t)] \\
&\quad + m(t)A \cos[2\pi(F_c \pm F_{LO})t]
\end{aligned}
\tag{4.8}
$$

Note that other non-linearities may also be employed with broadly similar results.

The first term contains components at baseband and at $2F_c$ which are not passed by the bandpass IF stage. The second term contains a d.c. component and a component at $2F_c$; these are similarly rejected by the IF filter. The final term contains *sum* and *difference* frequency components centred on $|F_c + F_{LO}|$ and on $|F_c - F_{LO}| = F_{IF}$, as obtained previously with a true multiplier.

Although we have considered explicitly only DSB–SC signals, the same results hold for other modulation schemes: the output from the IF stage is a modulated signal (DBS–SC, envelope AM, FM, and so on) with effective carrier frequency F_{IF}.

Adjacent Channel Rejection

The primary advantage of the superheterodyne principle is that it provides for high gain and selectivity in the IF amplifier. Adjacent radio channels close to F_c, say at $F_c \pm \Delta F$, give rise to components at the output of the mixer centred on $F_{IF} \pm \Delta F$. Provided the response of the IF amplifier is sufficiently small at frequencies $\pm \Delta F$ removed from the centre frequency F_{IF} these unwanted adjacent channel signals are strongly attenuated. The selectivity of the IF stage thus controls the *adjacent channel* rejection capability of the receiver. Since the IF amplifier operates at a fixed frequency band sophisticated multiple-tuned circuits or other highly selective filtering techniques can be readily incorporated. It is not necessary for the various sections of the IF amplifier to be able to *track* one another as the receiver is tuned since this is accomplished simply by adjusting the local oscillator such that the difference between the carrier frequency F_c and the local oscillator frequency F_{LO} corresponds to the fixed IF frequency F_{IF}.

Image Channel Rejection

Here it is noted that $|F_c - F_{LO}| = F_{IF}$ admits two possible values for F_c, namely

$$
\begin{aligned}
F_{c1} &= F_{LO} + F_{IF} \\
F_{c2} &= F_{LO} - F_{IF}
\end{aligned}
\tag{4.9}
$$

Hence, if the receiver is tuned to F_{c1} and there is also a radio signal with frequency F_{c2} present at the receiver input then both of these signals are converted down to the IF band where they are amplified and passed to the demodulator. The unwanted component is referred to as an *image signal* and the image rejection of the receiver depends on the RF response prior to the mixer. A tuned RF amplifier is often

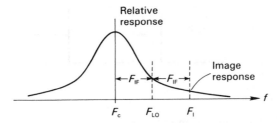

Fig. 4.4 Image response.

incorporated to provide a degree of image rejection. This amplifier has a centre frequency which is offset from the local oscillator by F_{IF} and is variable in step with the local oscillator by way of a ganged capacitor. The image frequency is given by

$$F_{image} = \begin{cases} F_c + 2F_{IF} & \text{if} & F_{LO} > F_c \\ F_c - 2F_{IF} & \text{if} & L_{LO} < F_c \end{cases} \qquad (4.10)$$

The problem of image rejection is illustrated in Fig. 4.4, from which it can be seen that the image response is reduced if the IF frequency is increased since F_{image} is then further removed from the main peak of the RF response.

A superheterodyne receiver, tuned to receive an AM broadcast signal at 1 MHz, employs a single tuned circuit RF amplifier with bandwidth $B_{RF} \equiv \pm 100$ kHz and an IF amplifier centred on 455 kHz. The local oscillator is set on the high side of the carrier. Determine (i) the local oscillator frequency; (ii) the image frequency; and (iii) the image rejection.

Worked Example 4.3

Solution:
(i) $F_{LO} = F_c + F_{IF} = 1$ MHz + 455 kHz = 1.455 MHz
(ii) $F_{image} = F_c + 2F_{IF} = 1$ MHz + 0.91 MHz = 1.91 MHz
(iii) From Equation 4.1 of Worked Example 4.1, the RF amplifier response is given approximately by

$$|H(f)|^2 = \frac{1}{1 + (f - F_c)^2/B^2}$$

$$|H(F_{image})|^2 = \frac{1}{1 + (910)^2/100^2} = 0.012$$
$$\equiv -19.2 \text{ dB}$$

The receiver provides some 19.2 dB of image rejection.

There are thus two conflicting requirements influencing the choice of IF frequency:
(i) It must be *low enough* so that, allowing for practical circuit elements with limited Q values, a sufficiently steep attenuation characteristic can be obtained outside the IF signal bandwidth to substantially suppress adjacent channel signals.

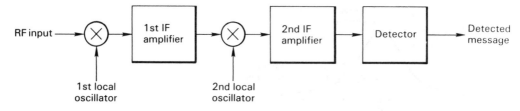

Fig. 4.5 Double-conversion receiver.

(ii) It must be *high enough* so that the RF amplifier provides adequate attenuation of unwanted signals at the image frequency.

The image rejection capability of a receiver can be further enhanced by using double conversion, as shown in Fig. 4.5. A relatively high IF frequency is chosen so that the RF amplifier provides substantial attenuation of image signals. Only modest adjacent channel rejection is provided, however, due to the limited Q of the filter components and the use of a high IF. The output from the first IF stage is then down-converted by a second mixer to a lower second intermediate frequency (second IF). This is chosen to be sufficiently low that good adjacent channel rejection can be obtained.

Automatic Frequency Control (AFC)

Frequency stability is especially important at high carrier frequencies since a small percentage frequency deviation corresponds to a very large absolute frequency deviation. Having said that, mixing at optical frequencies has been successfully demonstrated in research laboratories, and optical AFC loops are under development for optical fibre communication systems.

The frequency of a tunable local oscillator is liable to fluctuate slightly. This may result, for example, from temperature variations which may change the inductive and capacitive properties of the elements used in the frequency defining tuned circuits; the properties of the active devices (transistors) are also liable to change with temperature and so influence the local oscillator frequency. For FM systems frequency stability is especially important and an automatic frequency control (AFC) loop is usually incorporated to overcome these frequency drifts. The action of the loop is as follows: any error in the average frequency of the signal applied to the discriminator results in a d.c. offset at the output and the output can thus be lowpass filtered to remove signal components and extract the d.c. term. This can then be used to operate automatically on the local oscillator so as to minimize any frequency error.

Worked Example 4.4   A voltage controlled oscillator with a voltage-to-frequency conversion sensitivity of 200 kHz/V is found to drift by ± 100 kHz. It is to be used as the local oscillator for an FM radio receiver requiring frequency errors not exceeding ± 100 Hz. If the frequency discriminator produces 1 V output per 100 kHz of frequency offset determine the minimum voltage gain required in the AFC feedback path.

Solution: The AFC system may be modelled by the negative feedback system shown in Fig. 4.6, in which the local oscillator is assumed to operate above the carrier frequency. We make the following definitions:

F_c = input carrier frequency
F_{IF} = nominal IF frequency
F_E = local oscillator frequency error in the absence of AFC

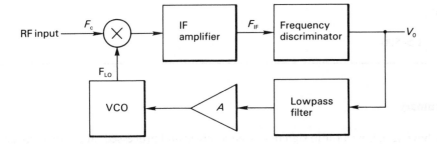

Fig. 4.6 Automatic frequency control loop.

F_{LO} = actual local oscillator frequency
K_F = voltage-to-frequency conversion gain of local oscillator [Hz/V]
K_V = frequency-to-voltage conversion gain of FM discriminator (V/Hz)
A = voltage gain of AFC feedback path

Now

$$F_{LO} = \underbrace{F_c + F_{IF} + F_E}_{\substack{\textit{local oscillator}\\ \textit{frequency in the}\\ \textit{absence of AFC}}} \underbrace{- K_F A V_o}_{\substack{\textit{correction term}\\ \textit{provided by}\\ \textit{AFC loop}}}$$

and

$$V_o = K_V \underset{\substack{\textit{actual frequency}\\ \textit{error}}}{(F_{LO} - F_c - F_{IF})}$$

$$= K_V F$$

Here $F = F_{LO} - F_c - F_{IF} = F_E - K_F A V_o$
       $= F_E - K_F K_V A F$
       $\Rightarrow F(1 + K_F K_V A) = F_E$

whence

$$F = \frac{F_E}{1 + K_F K_V A}$$

That is, the local oscillator drift is reduced by the *feedback factor* $(1 + K_F K_V A)$. Here

$K_F = 200 \text{ kHz/V} = 2 \times 10^5 \text{ [Hz/V]}$
$K_V = 1 \text{ V/100 kHz} = 10^{-5} \text{ [V/Hz]}$
$|F_E| < 100 \text{ kHz} = 10^5 \text{ [Hz]}$
$|F| < 100 \text{ Hz} = 10^2 \text{ [Hz]}$

Hence the following is required:

$$1 + K_F K_V A = 1 + 2 \times 10^5 \times 10^{-5} A > |F_E|/|F| = 10^3$$

that is,

$$A > 500$$

Summary

We have seen how a radio signal may be received and processed, allowing recovery of the message signal. The simple tuned radio frequency receiver principle was described and its limited adjacent channel capabilities noted. The superheterodyne receiver was then described and shown to offer considerable benefits in this respect, but in this instance image rejection must also be considered.

We noted that for effective adjacent and image channel rejection the local oscillator frequency and the intermediate frequency filter characteristics must be judiciously selected. It was suggested that double conversion, involving two IF stages, could usefully be adopted for demanding applications if the increased complexity (two local oscillators, two mixers and two IFs) is acceptable.

It was noted that automatic feedback loops may be required to correct for variations of signal level and for local oscillator frequency errors.

Problems

4.1 A TRF receiver uses an RF amplifier incorporating two identical tuned circuits prior to demodulation. Estimate the selectivity of this receiver, measured as the ratio of the -40 dB to -3 dB bandwidths.

4.2 A superheterodyne FM radio receiver uses a 10.6 MHz IF and, in the interests of economy, employs a single fixed tuned circuit RF amplifier prior to the mixer. The -3 dB bandwidth of this RF stage is adjusted to allow the receiver to operate over the band 88 MHz to 108 MHz. If the local oscillator operates above the carrier frequency, estimate

  (i) the frequency range over which the local oscillator must be tunable;

  (ii) the range of image frequencies the receiver encounters; and

  (iii) the worst case image rejection ratio, expressed in decibels.

Pulse Modulation Systems   5

☐ To describe the principle of pulse amplitude modulation (PAM).   **Objectives**
☐ To introduce the concept of sampling, show its relation to PAM and state
  the sampling theorem.
☐ To explain what is meant by aliassing distortion and how it can arise.
☐ To note briefly other pulse modulation schemes: pulse frequency modulation
  (PFM), pulse position modulation (PPM), pulse width modulation (PWM).
☐ To introduce the concept of time-division multiplexing (TDM).

Pulse Amplitude Modulation

In Chapter 3 we showed how a double sideband suppressed carrier modulated
signal could be produced as the product of a message signal $m(t)$ and a sinusiodal
carrier $x_c(t)$. The result was that copies of the message spectrum $M(f)$ were
produced at $\pm F_c$, i.e. in the vicinity of the spectral components of the carrier. We
now wish to examine the consequences of employing a pulse train, rather than a
sinusoid, as the carrier. This modulation scheme is shown in outline in Fig. 5.1.
The pulse train carrier $x_p(t)$ has the form

$$x_p(t) = \sum_k p(t - kT)$$

where

$$p(t) = \text{rect}(t/\tau) = \begin{cases} 1 & |t| < \tau/2 \\ 0 & \text{elsewhere} \end{cases} \tag{5.1}$$

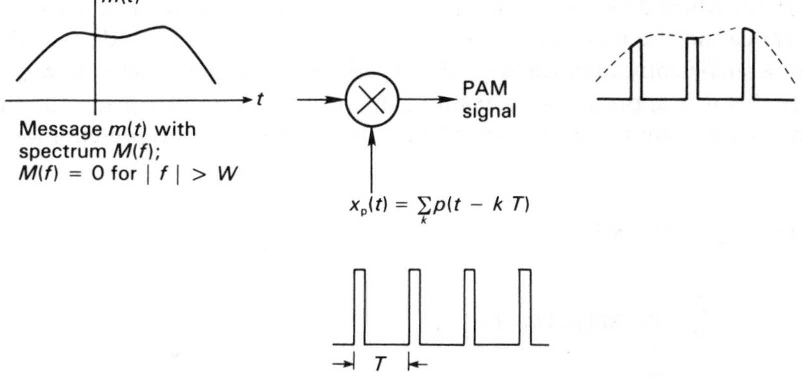

Fig. 5.1   Pulse amplitude modulation.

Recall the convolution theorem:
multiplication in the time domain
corresponds to convolution in
the frequency domain.

87

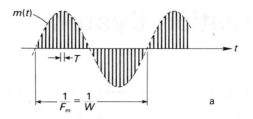

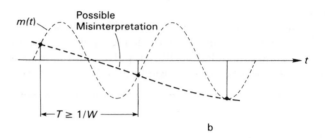

Fig. 5.2   Sampled sine waves. (a) $T \ll 1/W$; (b) $T \gtrsim 1/W$.

The pulse amplitude modulated (PAM) signal thus has the form

Note that this is the pulse carrier
analogue of DSB-SC.

$$x(t) = m(t)x_{\mathrm{p}}(t)$$
$$= m(t) \sum_k p(t - kT) \tag{5.2}$$

Before proceeding with the analysis, it is helpful to examine the form of $x(t)$ for various values of T. Recall that the message $m(t)$ is assumed strictly bandlimited to $|f| < W$, so that $m(t)$ changes only slightly over time intervals $\ll 1/W$. Hence, if $T \ll 1/W$, as in Fig. 5.2a, the heights of successive modulated pulses in $x(t)$ are almost equal and it is easy to 'guess' the likely form of $m(t)$ from the form of $x(t)$. Note that for $m(t) > 0$ the output pulses are positive and for $m(t) < 0$ the output pulses are negative. Thus it is expected that a close approximation to $m(t)$ should be obtained by lowpass filtering $x(t)$. This is indeed the case. Considering now Fig. 5.2b for which $T > 1/W$ it is not possible to guess $m(t)$ from $x(t)$; it transpires that $m(t)$ cannot be recovered from $x(t)$ in this case. We conclude that PAM requires a sufficiently high carrier pulse repetition frequency (PRF). In order to quantify this it is appropriate to examine the problem in the frequency domain. The analysis is commenced by considering the Fourier series representation for $x_{\mathrm{p}}(t)$:

$$x_{\mathrm{p}}(t) = \sum_k p(t - kT)$$

$$= \sum_{n=-\infty}^{\infty} c_n \exp(j2\pi nt/T)$$

$$= c_0 + \sum_{n=1}^{\infty} 2c_n \cos(2\pi nt/T) \tag{5.3}$$

where the last form arises since $x_p(t)$ is an even function which admits a Fourier cosine series expansion.

Hence,

$$x(t) = m(t)x_p(t)$$

$$= m(t)\left[c_0 + 2\sum_{n=1}^{\infty} c_n \cos(2\pi nt/T)\right]$$

$$= c_0 m(t) + 2c_1 m(t)\cos(2\pi t/T)$$
$$+ 2c_2 m(t) \cos(2\pi 2t/T)$$

$$\vdots$$

$$+ 2c_n m(t) \cos(2\pi nt/T) \tag{5.4}$$

A Fourier series representation for this repetitive pulse train was obtained in Chapter 2.

Examination of Equation 5.4 reveals that the PAM signal contains the following components:

$c_0 m(t)$ = the message, scaled by c_0
$2c_1 m(t) \cos(2\pi t/T)$ = DSB-SC signal with
(suppressed) carrier frequency = $1/T$
$2c_2 m(t) \cos(2\pi 2t/T)$ = DSB-SC signal with
(suppressed) carrier frequency = $2/T$
and so on. (5.5)

Since DSB-SC produces a scaled copy of the message spectrum at $\pm F_c$, it can be concluded that the PAM signal contains a multiplicity of scaled copies of the message at $\pm n/T$, n integer, as shown in Fig. 5.3. The relative *strengths* of these various copies depend on the Fourier coefficients c_n, which in turn depend on the

This is simply a consequence of the convolution theorem: multiplication of $m(t)$ by $x_p(t)$ to form $x(t)$ results in $X(f)$ corresponding to the convolution of $M(f)$ with the line spectrum $X_p(f)$.

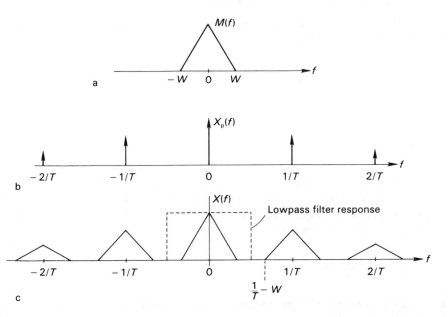

Fig. 5.3 Spectral regresentations of PAM. (a) Message spectrum, $M(f)$; (b) spectrum, $X_p(f)$, of pulse train $x_p(t)$, $T < 1/2W$; (c) PAM signal spectrum with $1/T > 2W$ showing recovery of $M(f)$ by lowpass filtering.

precise form of the pulse waveform, $p(t)$. The c_n are examined explicitly shortly; for the present simply note that provided $1/T > 2W$ the various replications of the message spectrum are disjoint in the frequency domain and the message can be recovered by lowpass filtering, as shown in Fig. 5.3.

Sampling

Sampling is a limiting case of PAM in which the pulses are of very short duration such that, for practical purposes, the PAM signal depends on the value of the message at only discrete points in time, $t = nT$. To represent this analytically it is convenient to consider the pulse $p(t)$ to take the form

$$p(t) = \frac{1}{\tau} \, \text{rect}(t/\tau) = \begin{cases} 1/\tau & |t| < \tau/2 \\ 0 & \text{elsewhere} \end{cases} \tag{5.6}$$

such that

$$x_p(t) = \frac{1}{\tau} \sum_k \text{rect}\left(\frac{t - kT}{\tau}\right)$$

$$= \sum_n c_n \exp(j2\pi nt/T) \tag{5.7}$$

whence

$$c_n = \frac{1}{T} \int_{-\tau/2}^{\tau/2} \frac{1}{\tau} \exp(-j2\pi nt/T) \, dt = \frac{1}{T}\left[\frac{\frac{1}{\tau}\exp(-j2\pi nt/T)}{-j2\pi n/T}\right]_{-\tau/2}^{\tau/2}$$

$$= \frac{1}{T} \frac{\sin(\pi n\tau/T)}{\pi n\tau/T} = \frac{1}{T} \, \text{sinc}\left(\frac{\pi n\tau}{T}\right) \tag{5.8}$$

Taking the limit as $\tau \to 0$, corresponding to short duration, large amplitude sampling pulses, we obtain

$$c_n = \frac{1}{T} \qquad \text{for all } n \tag{5.9}$$

since

$$\lim_{x \to 0} \text{sinc}(x) = 1$$

Hence with this idealization a sampled signal contains replications of the message spectrum $M(f) \leftrightarrow m(t)$ located at multiples of the sampling frequency $F_s = 1/T$. The replications are of equal strength corresponding to the message being scaled in amplitude by $1/T$. This may be written formally as

Again this follows readily from the convolution theorem. With $x_p(t) = \sum \delta(t - nT)$ we have $x(t) = m(t) \sum \delta(t - nT)$ giving

$$X(f) = M(f) * \frac{1}{T} \sum_n \delta\left(f - \frac{n}{T}\right)$$

$$= \frac{1}{T} \sum_n M\left(f - \frac{n}{T}\right)$$

$$X(f) = \frac{1}{T} \sum_n M\left(f - \frac{n}{T}\right) \tag{5.10}$$

The message is recoverable by lowpass filtering provided $F_s > 2W$, where W is the bandwidth of the (lowpass) message corresponding to the highest possible frequency component contained in the message. If $F_s < 2W$, spectral overlap occurs, as shown in Fig. 5.4, and lowpass filtering can recover only a distorted form of the

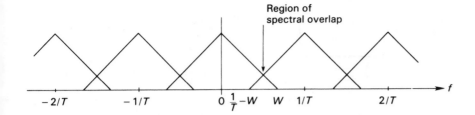

Fig. 5.4 Spectral overlap due to sampling at $1/T < 2W$.

message. The character of this form of distortion is best appreciated by reference to a sinusoidal message, as follows:
Consider

$$m(t) = \cos(2\pi F_m t) \qquad (5.11)$$

with spectrum

$$M(f) = \frac{1}{2}\,\delta(f \pm F_m) \qquad (5.12)$$

containing components at $\pm F_m$. If the signal is sampled at F_s such that $F_m = F_s/2 + \epsilon$ the resulting spectrum contains components at $nF_s \pm F_m$ as shown in Fig. 5.5. Notice that the output contains terms at $|f| = F_s/2 - \epsilon$ so that it is

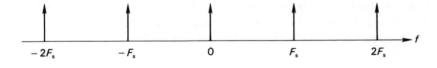

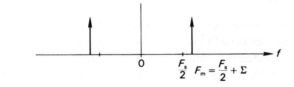

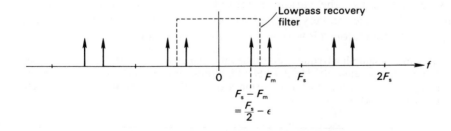

Fig. 5.5 Illustration of aliassing. (a) Spectrum of sampling pulse train, $F_s = 1/T$; (b) spectrum of cosinusoidal message; (c) spectrum of sampled signal with aliassed components at $f = \pm [(F_s/2) - \epsilon]$.

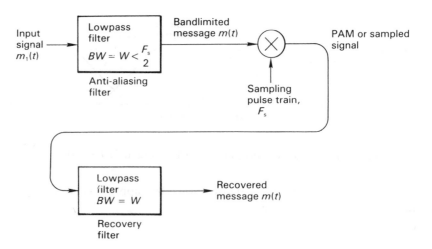

Fig. 5.6   A PAM or 'sampled data' system.

impossible for a lowpass filter to recover the original message terms at $\pm F_m = +(F_s/2 + \epsilon)$ without also passing the terms at $\pm(F_s/2 - \epsilon)$. The recovered signal is thus distorted. Note, however, that the output term at $(F_s/2 - \epsilon)$ could equally well be due to an input term at this frequency. If we assume the use of an ideal lowpass filter with bandwidth $B = F_s/2$ then the term $F_s/2 + \epsilon$ at the input is in a sense mass masquerading as a message component at $(F_s/2 - \epsilon)$. This is referred to as an *alias* or an *aliassed component* and this form of distortion is called *aliassing*. In order to minimize aliassing distortion the message can be passed through a sharp cut-off lowpass filter prior to sampling, as shown in Fig. 5.6. Since any practical filter has a finite transition region, the bandwidth of this *anti-aliassing filter* must be somewhat less than $F_s/2$.

Alias: *Name by which one is called on other occasions* (*Concise Oxford Dictionary*).

These observations are of such importance that we encapsulate them formally in a theorem.

The Sampling Theorem

There are more general forms of this theorem. See, for example, Cattermole, K.W., *Principles of Pulse Code Modulation*, Iliffe, 1969.

A strictly bandlimited signal $m(t)$ with spectrum $M(f) = 0$ for all $|f| > W$ may be recovered from samples $m(nT)$ taken uniformly in time at a rate $F_s = 1/T$ if and only if $F_s > 2W$. Provided this condition is satisfied then $m(t)$ may in principle be recovered from the sampled signal by using an ideal lowpass filter with bandwidth W.

Note that, unlike the various amplitude modulation schemes discussed in Chapter 3, message recovery from a PAM or sampled data signal format does not require a demodulator; a lowpass filter is all that is required.

Exercise 5.1   Suggest suitable sampling frequencies for the following signals: (i) telephone quality speech, bandlimited to the range 300 Hz–3.4 kHz; (ii) a music quality audio signal with spectrum extending to 15 kHz; and (iii) a television signal with an essentially lowpass spectrum extending to 5.5 MHz.

Worked Example 5.1   A telephone quality speech signal is generally considered to be concentrated in the band 300 Hz to 3.4 kHz but to allow for finite filter transition regions is often treated as having a spectral occupancy corresponding to 4 kHz lowpass. On these grounds a sampling frequency of 8 kHz is selected for a sampled data speech system and a *first-order* anti-aliassing filter with a −3 dB frequency of 3.4 kHz is employed. A similar filter is used for signal recovery. In testing the anti-aliassing performance it is found that if a 4.6 kHz sinusoidal signal is applied to the input an approximately sinusoidal signal at 3.4 kHz appears at the output. Explain why this is so and determine the relative strength of this component compared with a similar amplitude 'wanted' component at 1 kHz. Comment on the adequacy or otherwise of the anti-aliassing filter.

Solution: The first-order filter has a magnitude response given by

$$|H(f)| = \frac{1}{\sqrt{1 + (f/F_c)^2}} \tag{5.13}$$

where F_c = 3.4 kHz. Hence the 4.6 kHz input signal is attenuated by a factor of $[1 + (4.6/3.4)^2]^{-1/2} \simeq 0.59$ at the input to the sampler. This gives rise to a component in the sampled output of strength $0.59/T$ at 3.4 kHz. This is attenuated by a factor of $1/\sqrt{2}$ by the recovery filter to give an output signal of strength $\sim 0.415/T$ at 3.4 kHz. In contrast, the 1 kHz tone would be only slightly affected by the filters; attenuated by a factor of $[1 + (1/3.4)^2]^{-1/2}$ in each case equals $[1 + (1/3.4)^2]^{-1} \sim 1/T$ overall. The relatively high level of aliassing distortion encountered here suggests that a sharper cut-off anti-aliassing filter should be used.

Other Pulse Modulation Schemes

In Chapter 3 it was noted that a message could be impressed on to the amplitude, phase or frequency of a sinusoidal carrier, although the various amplitude modulation schemes were discussed in the most detail. Similarly here, while going into some detail on amplitude modulation of a pulse train carrier, the existence of other pulse modulation schemes broadly analogous to phase and frequency modulation are considered briefly. Of the various possibilities, perhaps the most widely known and employed are the following.

Pulse Frequency Modulation (PFM)

This is an almost exact analogue of frequency modulation (FM) of a sinusoidal carrier. The pulse repetition frequency of a pulse train is made to vary in sympathy with the message and signal recovery can be achieved as for FM. This modulation scheme is adopted for low cost, high performance analogue communication over optical fibre channels since it avoids any non-linearities in the opto-electronic devices impairing signal quality.

Pulse Position Modulation (PPM)

The position of each pulse within a time slot of duration T is made to vary in sympathy with the message. Signal recovery is achieved by measuring the variation of the time interval between the occurrence of a pulse and the occurrence of

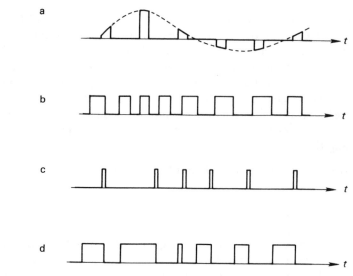

Fig. 5.7 Various pulse modulation schemes. (a) PAM, (b) PFM, (c) PPM, (d) PWM.

transitions in a regular timing signal at rate $F_s = 1/T$. The latter signal is extracted from the incoming modulated waveform, a narrowband filter or phase locked loop being used to smooth out perturbations due to the message.

Pulse Width Modulation (PWM)

The width or duration of the pulses is made to vary with the message; signal recovery can be achieved by lowpass filtering.

These various pulse modulation schemes are illustrated in Fig. 5.7 together with PAM for comparison purposes. There is a further class of pulse modulation schemes for analogue communication, namely *pulse code modulation* (PCM). However, these are quite different in character to those noted here and are of such importance that they are treated in some detail in the next chapter.

Time Division Multiplexing

It is convenient and appropriate here to introduce the idea of time division multiplexing (TDM) — the dual of frequency division multiplexing discussed in Chapter 3. For the present, attention is confined to TDM-PAM, time division multiplexing of pulse amplitude modulated signals.

Consider a set of N independent messages: $m_1(t)$, $m_2(t)$, ... $m_N(t)$, each strictly bandlimited to $|f| < W$. If these are sampled at $F_s = 1/T = 2W$ each results in a PAM signal with F_s samples/second. With the sampling for each message offset by a time interval T/N the various sampled signals can be added together to produce a composite *time-division multiplexed signal*. This is illustrated in Fig. 5.8 for $N = 4$. In this way several messages can be combined into a composite TDM-PAM format for transmission over a single communication channel at NF_s samples/second.

To recover the individual messages it is necessary to sample the composite TDM-

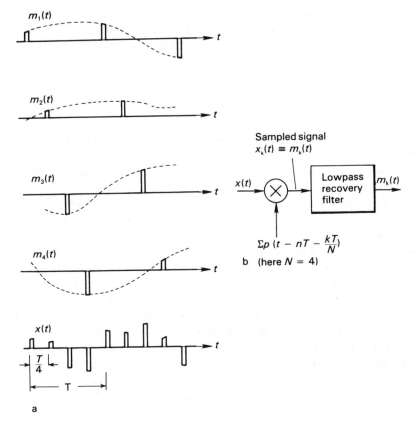

Fig. 5.8   Time division multiplexing of PAM signals. (a) TDM signal format; (b) demultiplexing and message recovery for kth channel.

PAM signal at appropriate epochs at a rate $F_s = 1/T$. For the example of Fig. 5.8, recovery of message $m_3(t)$ involves sampling $x(t)$ at times $t = nT + 3T/4$. The message is then recovered from this sampled signal by lowpass filtering.

The above discussion outlines the principle of TDM. It can be applied also to certain other analogue pulse modulation schemes, such as PPM, and is fundamental to many digital and data transmission systems. Perhaps the most widespread application of TDM, however, is to PCM telephony.

Summary

This chapter has been concerned with some of the more common methods of transmitting analogue signals via a pulse carrier. The sampling principle is at the heart of all such methods and this has been examined in both the time domain and the frequency domain. Using a frequency domain view we have seen that the sampling rate must be at least twice the frequency of the maximum message frequency component (i.e. twice the message bandwidth for a baseband signal) if the message is to be subsequently recoverable without distortion using a lowpass filter. This is the key result of the chapter.

Various modulation schemes involving variation of the amplitude (PAM), frequency (PFM), position (PPM) and width (PWM) of a pulse carrier have been described in outline. We saw also that several signals may be combined using pulse modulation and time-division multiplexing.

Problems

5.1 (a) Suggest suitable sampling frequencies for the following signals:
   (i)   A group of 12 speech signals, each of nominal bandwidth 4 kHz, which have been frequency division multiplexed (FDM) into the band 60 kHz to 108 kHz.
   (ii) A group of 5 FDM signals of the form in (i) which have been frequency division multiplexed into a minimum bandwidth equivalent lowpass signal.
   (iii) A composite FM stereo signal (to be discussed in detail in Chapter 8) which is effectively lowpass, extending from 30 Hz to 53 kHz.
   (b) State what bandwidth recovery filter is required in each of the above cases.

5.2 A sampling system uses a fourth-order anti-aliassing filter with a cut-off frequency of 10 kHz. The filter provides 24 dB attenuation per factor of 2 in frequency above cut-off (e.g. 48 dB at 40 kHz). A similar filter is used for signal recovery; the sampling rate is 30 kHz. On testing the system it is found that if a 1 V r.m.s. 25 kHz sine wave is applied to the input a low amplitude component at 5 kHz appears at the output. Estimate the relative strength of this interference term compared with the output obtained when a 1 V r.m.s. 5 kHz sine wave input is used.

5.3 (i) Thirty speech signals, each of nominal bandwidth 4 kHz, are to be combined using PAM-TDM. What is the effective sample rate for the combined signal?
(ii) If instead the 30 signals are combined using SSB-FDM to produce a minimum bandwidth lowpass composite signal and this signal is then sampled what is the minimum sampling rate required?

5.4 Use the convolution theorem to provide an alternative derivation for the spectrum of a PAM signal, as shown in Fig. 5.3.

Pulse Code Modulation 6

Objectives

- [ ] To describe and analyse the process of uniform quantization.
- [ ] To describe the principle of pulse code modulation (PCM) as a combination of quantizing, sampling and pulse encoding.
- [ ] To assess analytically the noise performance of a PCM system.
- [ ] To note briefly the merits of non-uniform quantization or companding for encoding certain signals, such as speech.
- [ ] To examine the application of PCM to television and to illustrate the subjective impairment of contouring.
- [ ] To discuss briefly differential pulse code modulation (DPCM) and delta modulation (DM) schemes.
- [ ] To outline the principles of, and note the CCITT recommentations for, PCM-TDM telephony.

The modulation schemes discussed so far provide for the representation of arbitrarily small perturbations in the message signal. In practical applications, however, a received signal is impaired by unwanted interference and noise and this limits the accuracy with which the message can be recovered. As an alternative, a certain small error in message representation can be accepted at the outset and the freedom this implies to obtain a degree of immunity to disturbance on the channel can be exploited. This is the essential idea behind pulse code modulation (PCM). Its practical implementation makes use of *sampling* together with two further processes: *quantization* and *pulse encoding*.

The invention of PCM is attributed to a British research engineer, Alec Harley Reeves. It was to some extent ahead of the technological developments required to make it a practical proposition. Only in recent years has it been applied internationally.

Quantization

It is appropriate first to consider quantization — the process whereby a continuous message signal $m(t)$ is replaced by an approximation $m_Q(t)$ assuming only discrete values. The process is illustrated schematically in Fig. 6.1a. The message is assumed to be bounded, $-\hat{V} < m(t) < \hat{V}$, and $m_Q(t)$ can assume any of 2^N possible output levels. The levels are taken to be uniformly spaced by an amount Δ, where

$$\Delta = 2\hat{V}/2^N \tag{6.1}$$

The *quantization error* $e(t)$ is defined as the difference between $m(t)$ and $m_Q(t)$:

$$e(t) = m(t) - m_Q(t) \tag{6.2}$$

and is bounded by $-\Delta/2 < e(t) < \Delta/2$, as illustrated in Fig. 6.1b. That is, $|e(t)| < \Delta/2$ and the worst case quantization error is thus $\Delta/2$. One means of

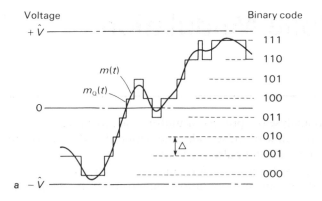

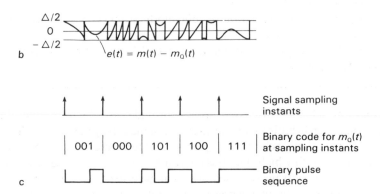

Fig. 6.1   Pulse code modulation. (a) Signal quantization and binary code assign-
ment; (b) quantization error wareforms; (c) sampling and pulse encoding.

specifying the fidelity of the quantized signal is by the ratio of the peak signal to the
worst case quantization error:

$$\frac{|\hat{m}|}{|\hat{e}|} = \frac{\hat{V}}{\Delta/2} \frac{\hat{V}}{(2\hat{V}/2^N)/2}$$
$$= 2^N$$

(6.3)

From this it can be concluded that the signal quality increases in proportion to the
number of levels, or is inversely proportional to the size of the quantel step, Δ.
While this is certainly a valid measure, it is not the most useful. After all, the
message $m(t)$ may reach its peak value only very occasionally while the quantiza-
tion error $|e(t)|$ reaches its peak value each time $m(t)$ crosses a boundary between
adjacent quantization levels. A more appropriate measure is obtained if the ratio
of the power in the signal to the power in the quantizing error is considered — the
signal-to-quantizing noise ratio SNR_Q. Considering the signal, power is proportional to voltage squared; hence

$$S \propto \overline{m^2} \triangleq \lim_{T \to \infty} \frac{1}{T} \int_{-T/2}^{T/2} m^2(t) \, dt$$

(6.4)

A peak-to-mean power ratio, α, can usefully be defined for the signal, usually expressed in dB as

$$\alpha_{\text{dB}} = 10 \log_{10} \left(\frac{\hat{V}^2}{\overline{m^2}} \right) \tag{6.5}$$

Considering now the quantization error $e(t)$, the waveform is closely approximated by triangular segments ranging over $\pm \Delta/2$. Hence $e(t)$ is uniformly distributed over the interval $(-\Delta/2, \Delta/2)$ and has mean square value given by

$$\overline{e^2} = \lim_{T \to \infty} \frac{1}{T} \int_{-T/2}^{T/2} e^2(t) \, \mathrm{d}t = \int_{-T/2}^{T/2} v^2 \, p_e(v) \, \mathrm{d}v \tag{6.6}$$

where $p_e(v)$ is the probability density function for the error $e(t)$. The latter is uniform on $(-\Delta/2, \Delta/2)$ of strength $1/\Delta$; hence

$$\overline{e^2} = \int_{-\Delta/2}^{\Delta/2} \frac{1}{\Delta} v^2 \, \mathrm{d}v = \frac{1}{\Delta} \left[\frac{v^3}{3} \right]_{-\Delta/2}^{\Delta/2} = \Delta^2/12 \tag{6.7}$$

It follows that

$$\text{SNR}_Q = \frac{\overline{m^2}}{\overline{e^2}} = \frac{12\,\overline{m^2}}{\Delta^2} = 3\,\frac{\overline{m^2}}{\hat{V}^2}\,2^{2N} = \frac{3 \times 2^{2N}}{\alpha} \tag{6.8}$$

Expressing this in dB:

$$\text{SNR}_Q \equiv 10 \log_{10}(3) + 20N \log_{10}(2) - \alpha_{\text{dB}}$$
$$= 4.77 + 6N - \alpha_{\text{dB}} \tag{6.9}$$

Note the equivalencing of *time averaging* and *statistical averaging*. This is often possible and can lead to considerable simplifications in analysis.

The signal-to-quantizing noise ratio improves 6 dB per extra bit allocated to the uniform quantizer.

A speech signal has a peak-to-mean power ratio of 10 dB. Ascertain the number of quantization levels $M = 2^N$, (a power of 2) required to ensure a signal-to-quantization noise ratio of at least 50 dB.

Worked Example 6.1

Solution:

Let $\quad \text{SNR}_Q \equiv 4.77 + 6N - \alpha_{\text{dB}} \geq 50$ dB
Then $\quad 6N \geq 50 + \alpha_{\text{dB}} - 4.77 = 55.23$
and $\quad N \geq 55.23/6 = 9.25$

Hence $N = 10$ is the smallest integer exceeding 9.25 and $M = 2^N = 2^{10} = 1024$ levels are required.

Sampling and Pulse Encoding

In the foregoing, it is assumed that the number of quantization levels is a power of 2. This allows each level to be represented by an N-bit binary number, as illustrated in Fig. 6.1c. By combining the operations of sampling, quantization and binary pulse encoding, pulse code modulation is obtained, in which quantized samples of the message are transmitted as binary pulse sequences, as shown in Fig. 6.1c. At the receiver, the binary signal is reconstructed into a quantized and sampled version of $m(t)$ and is then filtered to recover an approximation to $m(t)$. A major benefit offered by PCM is that binary encoding allows us to make use of digital transmission. As can be seen in a later chapter, this enables us to recover at the receiver, with arbitrarily small error, the binary signals. Hence the only significant signal impairment derives from the quantization process employed at the transmitter. Assuming there is no corruption of the binary signal in transmission the signal-to-noise ratio obtainable at the receiver is precisely the same as that developed previously in Equation 6.9. This deceptively simple statement requires some justification.

Consider a signal $m(t)$ bandlimited to $|f| < W$ and sampled at $F_s = 2W$. The message can in principle be recovered unimpaired by *ideal* lowpass filtering to $|f| < W$. If message quantization prior to sampling is now considered, the signal at the output of the quantizer may be written as

$$m_Q(t) = m(t) + e(t) \tag{6.10}$$

and hence the spectrum of the quantized signal has the form

$$M_Q(f) = M(f) + E(f) \tag{6.11}$$

Now $m(t)$ is bandlimited and can thus be sampled at $F_s = 2W$ without suffering aliassing distortion; not so $e(t)$, however! The error signal contains frequent, rapid transitions between $-\Delta/2$ and $\Delta/2$ and thus has considerable high frequency content. It can be concluded that $E(f)$ extends to $|f| \gg W$ and, after sampling, the various replications of $E(f)$ overlap such that an entire copy of $E(f)$ is *folded* into the interval. $(-F_s/2, F_s/2)$, as illustrated in Fig. 6.2. The result is that, due to aliassing of the quantization error, there is as much error power in the band $(-F_s/2, F_s/2)$ after sampling as was contained in the whole error signal $e(t) \Leftrightarrow E(f)$ prior to sampling. Hence on lowpass filtering the message signal $m(t)$ is recovered corrupted by an aliassed error signal with mean square value equal to $\overline{e^2}$.

This result is difficult to establish formally. See, for example, Cattermole, K.W. and O'Reilly, J.J., *Problems of Randomness in Communications Engineering*, Pentech Press, 1984.

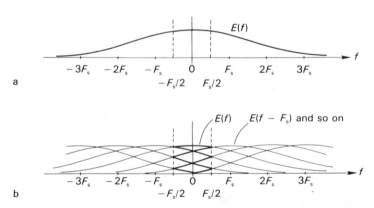

Fig. 6.2 Influence of sampling on quantization error spectrum. (a) Quantization error spectrum; (b) sampled quantization error spectrum with replications falling in the band $(-F_s/2, F_s/2)$.

The above argument assumes $W = F_s/2$. If the sampling rate is greater than twice the message bandwidth then the output error will be reduced by a further factor of $2W/F_s$ compared with $\overline{e^2}$.

Worked Example 6.2

A nominally 4 kHz speech signal is to be conveyed over a 64 kbit/s digital transmission system. Determine the signal-to-noise power ratio attainable assuming a 10 dB peak-to-mean power ratio for speech and uniformly spaced quantization levels.

Solution: The signal must be sampled at not less than 8 kHz to avoid aliassing distortion; hence we have available $64/8 = 8$ bits/sample. With minimum sampling rate and assuming an ideal recovery filter with bandwidth $W = 4$ kHz, the output signal-to-noise ratio is thus

$$
\begin{aligned}
SNR &= 4.77 + 6N - \alpha_{dB} \\
&= 4.77 + 48 - 10 \\
&\simeq 42.77 \text{ dB}
\end{aligned}
$$

Non-uniform Quantization

For PCM transmission of speech it is usual to employ a non-uniform series of quantization levels allowing small amplitude signals to be more finely quantized, as illustrated in Fig. 6.3. There are two reasons for this. First, the mean level of speech is much less than its peak value, the latter being attained only rarely. That is, there is a predominance of low level signals and these small fluctuations must be reproduced if the speech is to be intelligible. Second, different users of a system speak at different volume levels; with non-uniform quantization only the high signal level 'loud talkers' exercise the large amplitude quantel steps for which a high quantization noise results. Appropriate choice of the quantizing non-linearity can result in a nearly uniform signal-to-noise ratio for a wide range of talker volumes. Non-uniform quantization corresponds to an effective '*compression*' of the signal range which must be processed at the transmitter and makes use of a compensating 'expansion' at the receiver. The whole process, referred to as *companding*, has the effect of reducing the number of bits required for encoding. This is perhaps most easily appreciated with reference to a specific example. Study has

Various companding laws are discussed in some detail in Cattermole, K.W., *Principles of Pulse Code Modulation*, Iliffe, 1969.

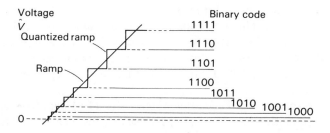

Fig. 6.3   Non-uniform quantizer (only positive half shown).

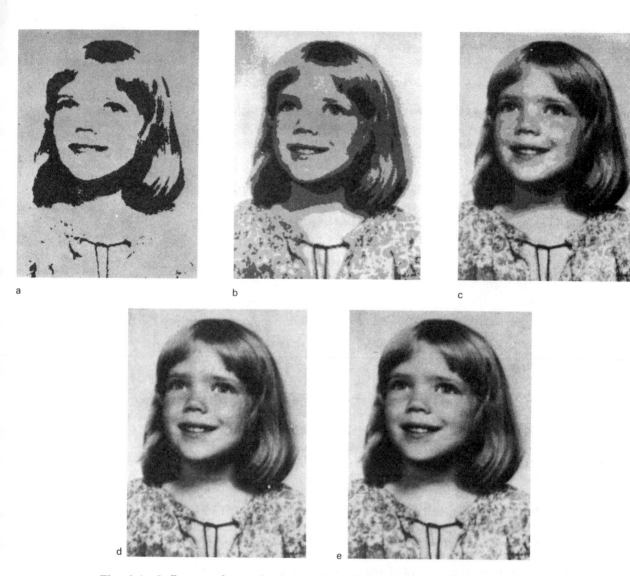

Fig. 6.4 Influence of quantization levels on PCM TV picture. (a) 1 bit/sample;
(b) 2 bits/sample; (c) 3 bits/sample; (d) 4 bits/sample; (e) 8 bits/sample.

shown that the range of significant fluctuation for a constant volume speech signal
is of the order of 30 dB while a further 30 dB variation can be encountered with different talker volumes. This demands a signal-to-smallest quantel step size of
60 dB, equivalent to 2000 levels or some 11 binary digits using uniform quantization. But uniform quantizing to this precision is quite unnecessary when dealing
with the higher-level signals and it transpires that an 8-bit representation suffices
with suitable companding.

In contrast to speech, PCM encoding of television (TV) signals generally makes
use of uniform quantization. This is in part because the mean signal level is more
closely defined but also has its origins in the way the eye perceives brightness variations. Uniform encoding to 7 or 8 bits has been found to provide adequate fidelity

for broadcast quality pictures. The influence of the number of bits on the subjective quality of the encoded TV picture is illustrated in Fig. 6.4. A subjective impairment known as *contouring* is clearly visible in the more coarsely quantized pictures.

Worked Example 6.3

A 5.5 MHz TV signal is to be PCM encoded into 8 bit samples using uniform quantization. Determine (i) the minimum sampling rate; (ii) the signalling rate (in bit/s) required for digital transmission; and (iii) the signal-to-noise ratio attainable.

Solution:
   (i) From the sampling theorem, a sampling rate of at least 11 MHz is required.
   (ii) Hence 8 bits/sample and 11 Msamples/second implies a signalling rate of 88 Mbit/s.
   (iii) The quality of a TV signal is often assessed in terms of the peak-to-peak signal-to-r.m.s. noise ratio. Hence, from Equation 6.9 with $\alpha = 1 = 0$ dB,

$$\text{SNR}_Q = 4.77 + 6N = 4.77 + 48 = 52.77 \text{ dB}$$

Differential Pulse Code Modulation

The preceeding example illustrates a major disadvantage of PCM — it can require very high digital transmission rates. One means of obviating this difficulty, which can provide a significant improvement if the signal changes only slightly from one sample to the next, is to encode only the difference between successive samples rather than the sample values themselves. This principle is illustrated in Fig. 6.5a. In practical implementations of this idea we encode not the differences between adjacent samples but the difference between the actual sample and our prediction on the basis of past behaviour, as shown in Fig. 6.5b. The receiver employs an identical predictor to the transmitter and uses the incoming samples both to correct and to update the present prediction.

A potential limitation of DPCM is that it suffers from rate limiting at large amplitude transitions. This can be minimized by the use of non-uniform quantization — by quantizing large differences more coarsely than small differences. DPCM has proved particularly effective as a means of reducing the bit rate required for digital transmission of TV signals. For example, Fig. 6.6 shows a TV picture encoded using 4 level (2 bits/sample) DPCM. Subjectively, the 2-bit DPCM picture is almost undiscernable from the 8-bit PCM picture shown in Fig. 6.4 and yet it requires only 1/4 of the digital transmission capacity. It is perhaps worth noting that the acceptability of DPCM encoded signals really does require subjective assessment and the conclusions depend significantly on the form of communication involved. With TV encoding, for example, we can exploit a phenomonon known as *spatial masking*; the eye is rather insensitive to coding errors in the presence of sudden large changes in brightness so coarse quantization of such changes is acceptable. Quite different conclusions can apply to speech or music signals for which spatial masking cannot apply.

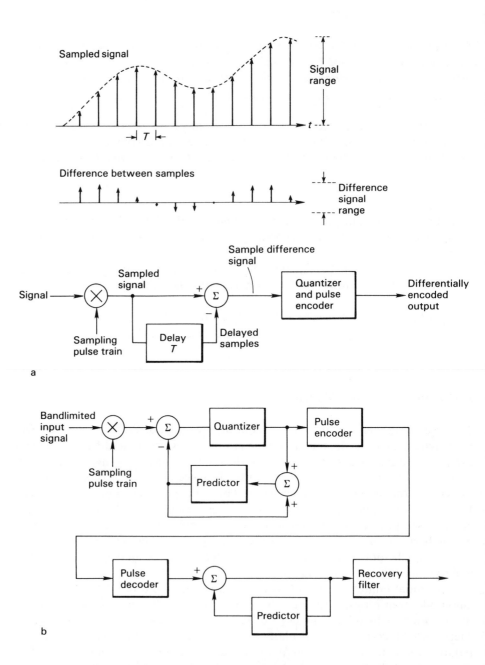

Fig. 6.5   Differential pulse code modulation. (a) Principle of differential encoding; (b) DPCM system using prediction from output.

Delta Modulation

A limiting case of DPCM is obtained if we represent signal changes by just one binary digit. This is known as *delta modulation* (DM). The main attraction of DM is the marked simplicity of the encoder and decoder, as shown in Fig. 6.7. The main application of DM is to low-cost encoding of speech signals. To avoid slope

Fig. 6.6   Four-level (2 bits sample) DPCM encoded TV picture.

overload, however, the sampling frequency must be high such that for similar signal-to-noise ratio performance the digital transmission rate required for DM can be broadly similar to that required for PCM. A variant of DM known as delta-sigma modulation (DSM) incorporates the integrator within the transmitter feedback loop. Among other advantages, this has the effect of enabling the transmission of d.c. components and further simplifies the receiver to just a lowpass filter.

See Steele, R., *Delta Modulation Systems*, Pentech Press, 1975, for a comprehensive discussion of this topic.

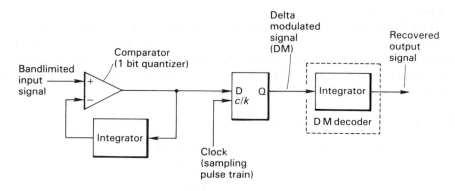

Fig. 6.7   Simple delta modulation system.

PCM-TDM Telephony

PCM is used in conjunction with TDM to realize multichannel digital telephony. The principle is shown in Fig. 6.8. While there are some variants, an internationally agreed CCITT standard provides for the combining of 30 speech channels, together with 2 subsidiary channels for signal/system monitoring, each speech

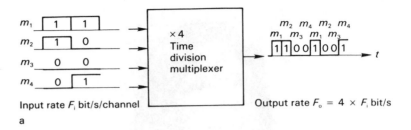

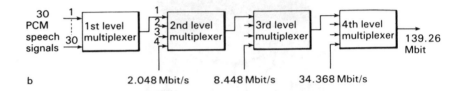

Fig. 6.8   Multichannel digital telephony: PCM-TDM. (a) Principle of PCM-TDM; (b) CCITT PCM-TDM hierarchy (note: higher levels, corresponding to ~280 Mbit/s, ~565 Mbit/s and above are under investigation/in use, but there is no formally agreed standard at the time of writing).

signal being sampled at 8 kHz and quantized into 8 bits. The digital rate per channel is thus 64 kbit/s and for the composite $30+2$ channel signal is 32×64 kbit/s = 2.048 Mbit/s. For convenience this is often referred to as a 2 Mbit/s PCM signal.

The CCITT also provides for higher levels of multiplexing combining groups of signals in blocks of 4. At each level the bit rate increases by slightly more than a factor of 4 since extra bits are added to provide for frame alignment, to facilitate satisfactory demultiplexing. The successive levels in the CCITT digital multiplex hierarchy are shown in Table 6.1.

CCITT (Comité Consultatif de Téléphone et de Télégraphie) — International Consultative Committee for Telephony and Telegraphy.

Table 6.1 CCITT PCM Hierachy

Level	No. 64-kbit/s channels	Approximate binary data rate
1	30	2 Mbit/s
2	120	8 Mbit/s
3	480	34 Mbit/s
4	1920	140 Mbit/s

Summary

In this chapter we have seen how analogue signals may be represented for transmission in digital form using pulse code modulation. There are three aspects to this: sampling, as discussed in Chapter 5; quantization of the samples such that only a finite discrete set of values are possible; and pulse encoding whereby each discrete value is represented by a unique digital pulse sequence. We have seen that the quantization process involves some impairment of the signal but that this can be small if a sufficiently large number of quantization levels are employed. Further, we noted that for some applications, such as speech encoding, companding (non-uniform quantization) may have merits.

For PCM television, companding is not generally used and the influence on picture quality of the number of bits employed to delimit the levels was illustrated. It was suggested that differential coding (DPCM) may reduce the number of bits required per sample. The limiting case of this is delta modulation in which only one bit per sample is employed; delta modulation schemes where described briefly.

Finally, the combined use of PCM and TDM techniques to allow several signals to be transmitted via a single digital link was described.

Problems

6.1 In contrast to speech, PCM encoding of television (TV) signals makes use of uniform quantization. Also, the quality of a TV signal is often assessed in terms of peak-to-peak signal-to-r.m.s. noise ratio. Consider a low bandwidth TV signal bandlimited to 3.2 MHz. If this is to be PCM encoded into 6-bit words using uniform quantization, determine (i) the minimum sampling rate; (ii) the signalling rate required for binary transmission; (iii) the signal-to-noise ratio attainable; and (iv) compare these requirements and performance figures with those appropriate to full 5.5 MHz bandwidth, 8-bit systems.

6.2 The digital audio compact optical disc (CD) system uses 16 bit quantization and a sampling rate of 44.1 kHz per channel. Assuming the audio signal has a peak to mean power ratio of 13 dB, occupies the frequency band 0 to 20 kHz and that the recovery filter has an *effective* bandwidth, allowing for the finite cut-off rate of a practical filter, of 22 kHz estimate the signal to quantization noise ratio attainable.

7 Digital Communications

Objectives

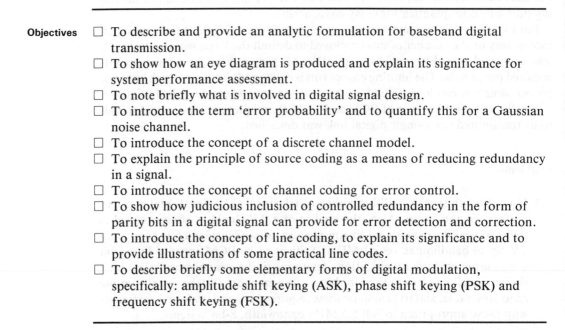

- [ ] To describe and provide an analytic formulation for baseband digital transmission.
- [ ] To show how an eye diagram is produced and explain its significance for system performance assessment.
- [ ] To note briefly what is involved in digital signal design.
- [ ] To introduce the term 'error probability' and to quantify this for a Gaussian noise channel.
- [ ] To introduce the concept of a discrete channel model.
- [ ] To explain the principle of source coding as a means of reducing redundancy in a signal.
- [ ] To introduce the concept of channel coding for error control.
- [ ] To show how judicious inclusion of controlled redundancy in the form of parity bits in a digital signal can provide for error detection and correction.
- [ ] To introduce the concept of line coding, to explain its significance and to provide illustrations of some practical line codes.
- [ ] To describe briefly some elementary forms of digital modulation, specifically: amplitude shift keying (ASK), phase shift keying (PSK) and frequency shift keying (FSK).

The general thrust of developments in modern communications is towards increasing digitization. For example, the widespread use of computers has led to an increased need for man–computer and computer–computer communications and thus an increasing proportion of the information we wish to transmit is inherently digital. Also, even in the field of human telecommunications exemplified by telephony, it is often found expedient to digitize the analogue signals (for instance, using PCM as discussed in the previous chapter) and to employ digital transmission. There are many reasons for this. For example, digital signals provide for virtually error-free transmission and such errors as do occur may be detected and/or corrected by using suitable data encoding and decoding. Also, digital signals are compatible with digital logic circuits and computer systems and can thus be economically processed and re-routed as required using such circuits and systems. Indeed, arguably it is the economic benefits of digital switching and signal processing which provide the main justification for the use of digital transmission for telephony; this admits the possibility of combining data, speech and other services into an *integrated services digital network* (ISDN), as noted briefly in Chapter 1. This point is returned to in due course; for the present various aspects of digital transmission are examined.

Two other texts in this series are concerned with digital logic techniques and computers — Stonham, T.J., *Digital Logic Techniques: principles and practice*, 2nd edition, 1986 and Downton, A.C., *Computers and Microprocessors: components and systems*, 2nd edition, 1987, both Van Nostrand Reinhold.

Digital Transmission

Consider a user of a computer terminal equipped with a typewriter-like keyboard.

Each time the user strikes a key a binary code is generated which identifies this key. For example, the numeral 9 might be represented by the 8-bit binary word 10001001. This code must be transmitted to the computer. If the computer is local to the user it may be most convenient to send the signal along 8 parallel wires; this might apply, for example, to the keyboard of a personal computer. This mode of operation is referred to as bit-parallel, word-serial. If the code 00000000 represents no key then the computer can recognize that a key has been struck when one or more bits take the value 1.

For transmission over greater distances the above scheme is inconvenient since it involves the use of 8 separate signal wires together with a 9th wire to provide a reference (ground). A bulky and expensive cable would thus be required. To circumvent this difficulty the data can be *serialized* for transmission over a single pair of wires, as shown in Fig. 7.1. A start bit can be added to the beginning of each character code to give the receiver warning that data are about to be sent and one or more stop bits are appended to delimit the end of a character. The receiver then interprets the binary sequence with reference to a *clock oscillator*. This is referred to as *asynchronous* serial transmission since a character can appear at any time and the receiver interprets the data accordingly. It is an appropriate basis for man–machine communication in which there are likely to be significant, variable, pauses between successive characters. It is quite common for a return channel to be

Special purpose integrated circuits are available to ease asynchronous communications: ACIA — Asynchronous Communication Interface Adaptor.

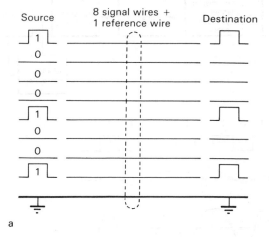

a

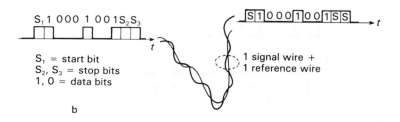

S_1 = start bit
S_2, S_3 = stop bits
1, 0 = data bits

b

Fig. 7.1   (a) Bit-parallel, word-serial data transmission; (b) asynchronous-serial data transmission.

provided, generally using the same pair of wires, and for the computer to echo the character to the display screen of the terminal. In this way the user obtains visual confirmation that the computer has received the transmitted data uncorrupted.

For machine–machine communication, however, this mode of operation is necessarily wasteful. There is no reason for a large amount of data not to be sent as a continuous binary digit stream, as a sequence of positive and negative pulses for example, provided the transmitter and receiver can be kept in step. This is known as synchronous transmission. It provides for efficient digital communication and is the technique adopted for the transmission of digitized speech samples in PCM systems. Accordingly we will discuss this method in some detail.

Baseband binary transmission principles

Recall that loss and distortion characteristics for various channels were discussed briefly in Chapter 1.

A binary transmission system is shown in outline in Fig. 7.2. The transmitted signal, when it arrives at the output of the channel, is attenuated and distorted due to the loss and restricted bandwidth of the channel. It is the role of the receiver to recover a faithful reproduction of the original data. This is made all the more difficult by the fact that the much attenuated signal is corrupted by random fluctuations, referred to as noise. The high frequency attenuation of the channel is

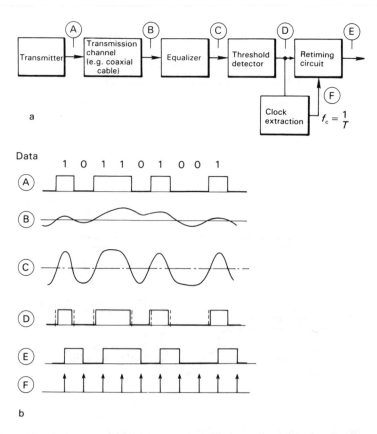

Fig. 7.2   Binary digital transmission system. (a) System block diagram; (b) waveforms in the absence of noise.

110

partially corrected by use of an equaliser — a frequency dependent network which enhances high frequency components. This equalised waveform is then applied to a threshold device which restores the signal to its original form except that the timing of the transitions may be irregular. This arises because the transition times depend on the precise form of the channel output, which will depend on the adjustment of the equaliser frequency response relative to the channel. Also, of course, the signal at the input to the threshold circuit is corrupted by noise and this too can perturb transition timings. Hence, to restore the relative regularity of data transitions a clock waveform of frequency $F_c = 1/T$ is extracted from the signal and used to retime the data. This combination of threshold and retiming circuitry provides for data recovery and corresponds to making decisions on the basis of samples of the signal taken at regularly spaced intervals of T seconds, the sampling epochs being adjusted to provide data extraction at the peaks and troughs of the equalised signal waveform. This provides maximum immunity to noise since at these points a noise voltage in excess of the difference between the threshold and the peak/trough signal value will need to occur before a decision-error is induced. Given this overview, we now proceed to a more detailed examination of the problem.

This clock recovery process is not discussed here but a relatively simple treatment is available in Cattermole, K.W. and O'Reilly, J.J., *Problems of Randomness in Communication Engineering*, Pentech Press, 1984.

Intersymbol Interference

Consider a binary signal, $x(t)$, given by

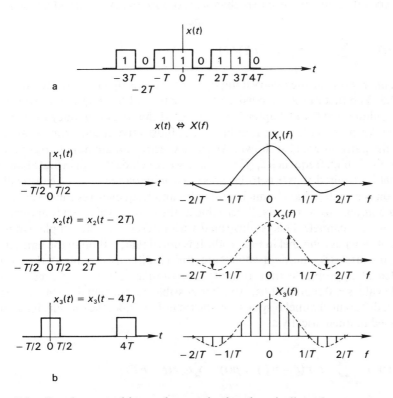

Fig. 7.3 Synchronous binary data and related periodic pulse patterns.
(a) Synchronous binary data signal, $1/T$ bit/s; (b) spectrum of a rectangular pulse and of related periodic pulse patterns.

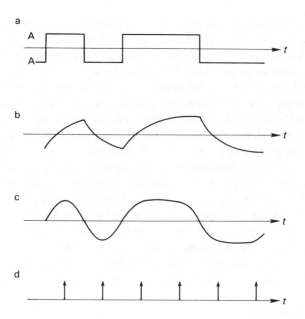

Fig. 7.4 Effect of bandlimiting on a binary data signal. (a) Rectangular data sequence; (b) response of first-order lowpass channel to (a); (c) zero intersymbol interference (isi) channel response to (a); (d) nominal decision times.

$$x(t) = \sum_{n = -\infty}^{\infty} a_n \, \text{rect}((t - nT)/T) \tag{7.1}$$

where $\{a_n\}$ represents the binary data, $a_n = 0$ or 1. This signal has the form shown in Fig. 7.3. The first question to ask is: 'What bandwidth is required to transmit the signal unimpaired?' In Chapter 2 we suggested that the occupancy of an isolated rectangular pulse has the form in Fig. 7.3b; illustrative spectra for certain repetitive pulse patterns are also shown. All these spectra contain non-zero components at arbitrarily high frequencies, $|f| \to \infty$, and we conclude that an infinite bandwidth channel is required if $x(t)$ is to be preserved unimpaired. More realistically, the question is asked: 'what channel bandwidth and frequency response is required if the data $\{a_n\}$ are to be readily discernable at the output when $x(t)$ is applied to the input of the channel? This is illustrated with reference to the simple, single time constant, lowpass channel in Fig. 7.4b. It is noted that, although the limited bandwidth distorts $x(t)$, the data values are still readily discernable if the output is sampled at appropriate time instants and sample values > 0 represent 1s and sample values < 0 represent 0s. Another possible output signal form, corresponding to a different channel response, is shown in Fig. 7.4c; the output signal may be expressed mathematically as

$$y(t) = \sum_{n = -\infty}^{\infty} a_n \, p(t - nT) = p(t) * \sum_n a_n \, \delta(t - nT) \tag{7.2}$$

where $p(t)$ is the signal element pulse shape corresponding to the response of the channel to an isolated pulse. Examination of Fig. 7.4b reveals that the actual values of the samples corresponding, for example, to a 1 can vary widely. It is seen

Note the use of the $*$ notation to represent a digital signal as the convolution of a basic pulse element waveform $p(t)$ with a data sequence comprised of impulses.

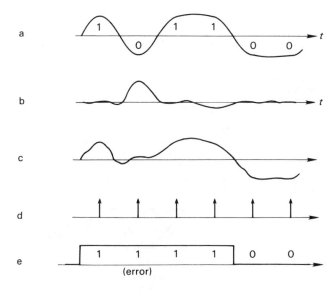

Fig. 7.5 Influence of noise and interference on binary transmission. (a) Signal, no noise; (b) notional noise waveform; (c) signal and noise; (d) decision times; (e) recovered data.

shortly that this may be undesirable since the closer the signal is to zero at the sample time the more likely noise and interference on the channel is to cause an error, as illustrated in Fig. 7.5. In the absence of the latter disturbances, variation in the signal sample values corresponding to 1s, and similarly 0s, is caused by the presence or absence of data pulses in adjacent time slots. If $a_n = 1$ and $a_{n+1} = 0$, a slowly decaying channel response of the form of Fig. 7.4b results in the output at time $t = (n + 1)T$ being more positive than if $a_n = 0$. It is said that the various symbols are interfering with one another and this phenomenon is referred to as *intersymbol interference* (ISI).

Some practical channels exhibit very severe ISI characteristics with the interference extending over many adjacent bit intervals.

The Eye Diagram

ISI is most readily observable if the synchronous superposition of all possible data patterns is considered, as shown schematically in Fig. 7.6a. This effect is achieved in the laboratory by triggering the oscilloscope on the data clock signal so that the digital signal becomes overlapped at multiples of the clock interval; Fig. 7.6b provides an example. The resulting pattern is referred to as an *eye diagram* in view of the shape of the central section. The eye is open if there is little ISI in which case all data 1 values are nearly equal at the sample time, and similarly for 0s. The eye diagram thus provides a ready visual check in the time domain of the adequacy of the transmission channel bandwidth/frequency response.

Signal Design

ISI depends on the basic signal element pulse shape, $p(t)$. If it is desired to have strictly zero ISI then $p(t)$ must satisfy

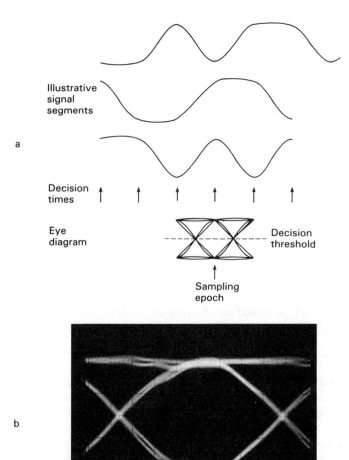

Illustrative
signal
segments

a

Decision
times

Eye
diagram

Decision
threshold

Sampling
epoch

b

Eye diagrams are also used to
assess multi-level systems. For
m-level signalling $(m-1)$ 'eyes'
occur vertically one above
another.

Fig. 7.6   Binary eye diagrams. (a) Formulation of eye diagram by superposition of data segments. (b) Observed eye diagram for an experimental binary transmission system.

$$p(0) = k \qquad \text{a non-zero constant}$$
$$p(nT) = 0 \qquad n \neq 0 \tag{7.3}$$

Fig. 7.7a provides an illustration of pulses satisfying these constraints. It is in principle possible to obtain zero ISI and yet have a limited bandwidth. To illustrate this recall that the spectrum of a rectangular pulse, rect(t), has the form sinc(f) $\triangleq \sin(\pi f)/\pi f$. As a consequence of the near symmetry of the Fourier transform relations (Equations 2.36a, b) which relate time domain pulse signals and their frequency domain representations, it transpires that a sinc(t) pulse has a rectangular, strictly bandlimited, spectrum. This result, established formally in Equation 2.46, is illustrated in Fig. 7.7b. Hence if we could realize a channel with a sinc(t/T) pulse response it would provide zero ISI while having a bandwidth of only $1/2T$, one-half of the signalling rate. Further, it can be shown that this is the limit

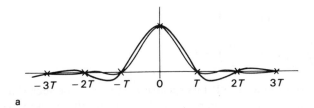

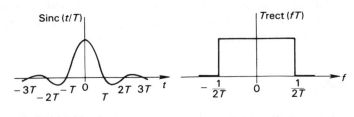

Sinc (t/T)

Trect (fT)

A specific zero isi signal

Corresponding bandlimited spectrum

b

Fig. 7.7   Illustrative zero isi signals. (a) Zero isi signals constraints marked ×;
(b) sinc (t/T) and its spectrum.

case; zero ISI signalling requires a bandwidth of not less than $1/2T$. Interesting and important though this result may be, it should not be interpreted too enthusiastically. It is not being suggested that strictly zero ISI can be obtained in practice — only that it is possible *in principle* in a restricted bandwidth. Note that the function sinc(t/T) extends over $-\infty < t < \infty$. A practical, causal, channel pulse response must be zero for $t < 0$ and even if a delayed version is considered, sinc$[(t - \tau)/T]$, there is no finite value τ such that the signal is causal, i.e. strictly zero for $t < 0$. Hence a more meaningful conclusion is that a signal $p(t)$ which is a close approximation to sinc$[(t - \tau)/T]$ may provide almost zero ISI and yet have a limited bandwidth. There are many other functions which provide zero ISI and although these require a bandwidth greater than $1/2T$ it is generally expedient to choose such a function, with a less steeply bandlimited spectrum, as the target for practical approximation. This reduces the complexity of the filter required and increases the tolerance of the system to variations in the transmission rate.

It is not uncommon to consider non-causal signals for the purpose of analysis in communication engineering. These may often be rendered approximately causal by introducing an appropriate delay.

Zero ISI Signals

The constraints on $p(t)$ to ensure zero ISI, represented by Equation 7.3, may be stated compactly as

$$p(t) \sum_n \delta(t - nT) = \delta(t) \tag{7.4}$$

That is, while $\sum_n \delta(t - nT)$ contains impulses at times $t = n/T$, multiplication by a zero ISI $p(t)$ anihilates all except the impulse at $t = 0$ since $p(t)$ is zero at $t = n/T$, $n \neq 0$. The corresponding constraints in the frequency domain on $P(f)$, the spectrum of $p(t)$, are obtained by Fourier transforming Equation 7.4

Note once more: multiplication
in the time domain gives rise to
convolution in the frequency
domain.

$$P(f) * \frac{1}{T} \sum_n \delta \left(f - \frac{n}{T} \right) = 1$$

$$\Rightarrow \frac{1}{T} \sum_n P \left(f - \frac{n}{T} \right) = 1 \tag{7.5}$$

Hence for $p(t)$ to provide zero ISI its spectrum $P(f)$ must be such that if replicated along the frequency axis at intervals of $1/T$ the resultant sums to a constant. Clearly the ideal lowpass function of Fig. 7.7 satisfies this constraint, as do many other functions. Note in particular though that any $P(f)$ with bandwidth less than $1/2T$ clearly cannot provide zero ISI since then there would be gaps in the replicated spectrum. We thus term the ideal lowpass function of Fig. 7.7 the *minimum bandwidth* characteristic for zero ISI signalling.

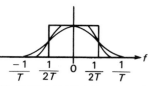

Illustrative spectra corresponding
to zero ISI signals.

Error Probability

The relative immunity to noise provided by a properly designed digital system has been alluded to on a number of occasions; this is now examined quantitatively. Recall the system of Fig. 7.2a in which a binary signal, taking values $\pm A$ at the decision instants, is corrupted by additive noise. The noise is assumed to be Gaussian, such that samples of the noise voltage are distributed according to

$$p(v) = \frac{1}{\sqrt{(2\pi)}\sigma} \exp(-v^2/2\sigma^2) \tag{7.6}$$

where σ is the r.m.s. noise voltage. This noise appears at the input to the decision circuit added to the signal voltage, which is $+A$ for a 1 and $-A$ for a 0, at which point the signal then has a mean value $+A$ when a 1 is transmitted and $-A$ when a 0 is transmitted. These voltage distributions for signal-plus-noise are shown in

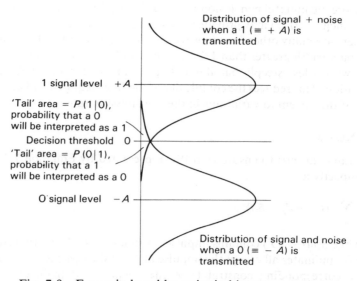

Fig. 7.8   Errors induced by noise in binary transmission.

Fig. 7.8. Errors occur when a transmitted 1 is interpreted at the receiver as a 0 and vice versa. These events occur with probability $P(0|1)$, $P(1|0)$ given by

$$P(0|1) = \int_{-\infty}^{0} \frac{1}{\sqrt{(2\pi)}\sigma} \exp[-(v - A)^2/2\sigma^2]\, dv \tag{7.7a}$$

and

$$P(1|0) = \int_{0}^{\infty} \frac{1}{\sqrt{(2\pi)}\sigma} \exp[-(v + A)^2/2\sigma^2]\, dv \tag{7.7b}$$

The average bit error probability is given by

$$P_e = P(1)P(0|1) + P(0)P(1|0) \tag{7.8}$$

where $P(0)$, $P(1)$ are the element probabilities — the relative frequencies of occurrence of 0 and 1 respectively. Usually it is assumed that 1s and 0s are equally likely in which case $P(0) = P(1) = 1/2$. Using the change of variable $x = (v - A)/\sigma$ in Equation 7.7a and $x = (v + A)/\sigma$ in Equation 7.7b:

$$P(0|1) = P(1|0) = \int_{A/\sigma}^{\infty} \frac{1}{\sqrt{(2\pi)}} \exp[(-x^2/2)\, dx \triangleq T(A/\sigma) \tag{7.9}$$

whence:

$$
\begin{aligned}
P_e &= P(0|1)/2 + P(1|0)/2 \\
&= T(A/\sigma)/2 + T(A/\sigma)/2 \\
&= T(A/\sigma) \tag{7.10}
\end{aligned}
$$

This is known as the Gaussian tail function. It is plotted in Fig. 7.9 as a function of the signal-to-noise ratio A/σ expressed in dB:

$$\text{SNR}_{dB} = 20 \log_{10}(A/\sigma) \tag{7.11}$$

Note that the error probability falls very rapidly with increasing signal-to-noise ratio. For example, with $A/\sigma \simeq 6 \equiv \text{SNR}_{dB} \simeq 15.5$ dB, $P_e \simeq 10^{-9}$. This is a realistic design figure for a high quality digital transmission system in a telecommunications network and at this level a 1 dB increase in signal-to-noise ratio reduces the error probability by more than two orders of magnitude. To appreciate just how low these figures are, consider the problem of measuring the error probability for a practical system. If a fixed, pseudo-random, data pattern which is known in advance to the receiver is transmitted then any errors which occur can be noted. To have reasonable confidence in the results an average must be taken. Assume that this is done by recording N, the total number of bits transmitted from some arbitrary start point until the time that 100 errors are detected. The average error probability can then be estimated as $100/N$. For $P_e \simeq 10^{-9}$ this implies $N \simeq 10^{11}$ bits. At a data rate of 2 Mbit/s this would require a measurement time of the order of $10^{11}/(2 \times 10^6) \simeq 50\,000$ seconds; approximately 14 hours!

Discrete Channel Models

In the foregoing we have discussed how digital signals may be transmitted over what is essentially an *analogue* communication link and may then be recovered at

The notation $P(0|1)$ denotes conditional probability. This is the probability that a 0 is detected when a 1 is present at the input to the receiver.

In this case the 1s and 0s are said to have equal *a priori* element probabilities.

Note that $\text{SNR}_{dB} = 20 \log_{10}(A/\sigma)$ since A and σ are voltages.

A pseudo-random binary sequence is readily generated using a maximal length feedback shift register or *m-sequence* generator — a particular form of sequential logic circuit. The binary *m*-sequence produced is periodic with pattern length equal to $2^N - 1$ bits for an N-stage shift register. Specific sequence generators recommended by CCITT have $N = 15$ and $N = 23$ so the test patterns are extremely long!

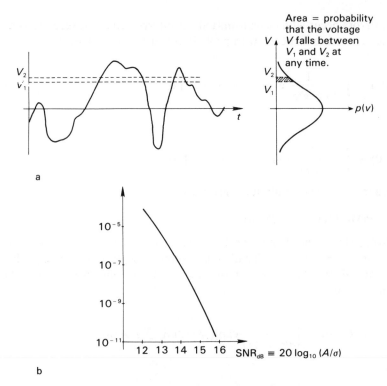

Fig. 7.9   Error probability for binary transmission in the presence of Gaussian noise. (a) Interpretation of probability density function for a noise waveform; (b) Gaussian tail probability: P_e versus SNR_{dB}.

the receiver, possibly with the occasional digit being in error. The main emphasis has been on binary transmission. With this perspective simplified, discrete, channel models for digital transmission systems, as shown in Fig. 7.10, can be introduced. The digital channel is seen to correspond to a *probabilistic* mapping between input and output digits. For binary transmission, an input 1 has a probability $P(0|1)$ of appearing incorrectly at the output as a 0 and probability $1 - P(0|1)$ of appearing correctly as a 1. Similarly for an input 0 there is a probability $P(1|0)$ that it appears at the output as a 1, and so on. If $P(0|1) = P(1|0)$ the channel is symmetrical in the way it treats 1s and 0s; this widely adopted channel model is referred to as the *Binary Symmetric Channel* (BSC) and is shown in Fig. 7.10a. Multi-level, or *m*-ary, signalling can also be employed in which case it is

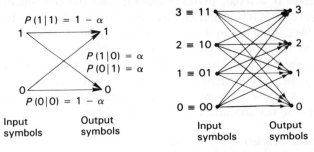

Fig. 7.10   Discrete channel representations. (a) Binary symmetric channel (BSC); (b) discrete quaternary channel.

possible for a signal element or symbol to convey more than one bit of information. For example, if there are four possible levels then each can be given a binary designation 00, 01, 10 or 11 and we can send 2 bits per symbol. A corresponding discrete channel representation is shown in Fig. 7.10b.

An m-level system can convey $\log_2 m$ bits per symbol since $N = \log_2 m \Rightarrow m = 2^N$ so that each level can be identified by an N-bit binary number. If m is not a power of 2 this result still applies but the data must be coded on to a *sequence* of symbols to achieve this *information capacity* per symbol on average.

Coding for Digital Transmission

Having shown how real, analogue, communication links can be used to provide purely digital channels attention can now be turned to the problem of transmitting data over such a channel and consequently to the subject of coding for digital transmission.

Source Coding

Source coding is concerned with the following problem: given a source of information how should messages from this source be represented such that on average the information is conveyed using the minimum number of bits. It is not attempted to treat this problem formally but rather to demonstrate the principle with reference to a specific example: that of encoding English text.

Messages are constructed as sequences of symbols, the set of possible symbols being known as the source *alphabet*. In this case the source alphabet is composed of the 26 letters A–Z, the numerals 0–9, a space and various punctuation characters. It may also be necessary to distinguish between upper case and lower case letters. The total source alphabet is thus considerably larger than the usual 26 letters and if it is desired to represent each symbol by a unique, fixed length, binary code some 7 bits are required giving $2^7 = 128$ distinct possible characters. A 7 bit code widely used for this purpose in computer and data communications, known as ASCII (American Standard Code for Information Interchange), is illustrated in Table 7.1.

Table 7.1   Seven-bit ASCII Code

Character	Binary code	Character	Binary Code
Communication control characters	0000000 — 0011111	A	1000001
		Z	1011010
Space	0100000	[	1011011 —
!	0100001	Punctuation and symbols	
Punctuation and symbols		\	1100000
/	0101111	a	1100001
0	0110000		
		z	1111010
9	0111001	{	1111011
		Punctuation and symbols	
:	0111010		
Punctuation and symbols		~	1111110
@	1000000	Delete	1111111

119

Here all characters are denoted by code words of the same length. An alternative strategy is to allocate short codes to characters which occur most often and long codes to characters occurring only infrequently. In this way the average number of bits required to transmit a message can be minimized. A familiar example of this latter strategy is provided by the Morse code. This was designed on the basis of the observed relative frequencies of letters in English text and is based on the signalling elements dot, dash and pause. Since in English the letter E occurs very frequently it is allocated a very short code, a single dot. On the other hand, the letter Y is rather infrequent and accordingly attracts a relatively long code. A disadvantage of this kind of code is that the signalling time varies from character to character (2 units for E, 13 units for Y, and so on). One consequence of this is the need to introduce a specific character to separate letters, as well as the usual space character to separate words, although some variable length codes avoid this problem. More seriously, the variable character length is inconvenient for automatic data processing. For example, if it is necessary to refer to the 12th member of a string of ASCII characters we can readily do so. In contrast, to locate the 12th member of a string of Morse coded characters each character must be read in turn, since only by interpreting one character do we find out where the next one starts. Hence, if data is to be not only transmitted but also stored and processed a variable length code may be inappropriate. On the other hand, if the primary concern is with serial transmission then efficient source encoding can be worthwhile. For machine–machine communication a simple regular code, such as ASCII, has many attractions. For human–human communications different considerations can apply and the Morse code is widely adopted for amature radio and for ship–ship/ship–shore communications. Generally this is based on hand signalling in which operators tap out the letter codes by hand using a key switch to gate on and off an oscillator. Some illustrative Morse code character representations are shown in Table 7.2.

Table 7.2 Illustrative Morse Code Representations

Letter	Code
E	.
A	.−
Y	−.−−

If it is required to transmit text yet more efficiently account is taken not only of individual letter frequencies but also of the relative frequencies of letter groups. For example, since in English there are no words in which the letter Q occurs without it being followed immediately by a U, the U is in a sense redundant: it carries no *information* when following a Q. The likelihood of a U occurring is increased to the level of certainty given that a Q has just appeared; it is said that there is strong *conditionality* or that the symbols are *correlated*. This is a very special case, but there are many less extreme examples. The probability of occurrence of each member of the source alphabet is, in general, strongly influenced by what has gone before. Very efficient encoding of text can be achieved if these symbol correlations are considered.

This has been discussed at some length not because of the practical importance of this particular example, which is slight, but because it provides a useful illustration of the principle of source coding. Also, only discrete sources have been considered: sources characterized by a countable alphabet. The same general ideas apply more broadly, however. Indeed, there is a sense in which the DPCM schemes of the previous chapter constitute source coding. The correlation between adjacent PCM samples is exploited by using a shorter, albeit fixed length, code to represent the change in the sample values rather than their absolute value. Source coding is a very active field for present day research and there are many subtleties, alternative strategies, even competing philosophies. We will not pursue these further here but simply note that whatever the efficiency of the source code employed the result is a sequence of binary symbols. Attention can now be turned to the problem of conveying these symbols with as few errors as possible over a discrete communication channel.

Error Control Coding

The presence of noise and interference on a communication link can result in corruption such that the received signal may differ from that which is sent. For a digital communication system this means that the received data pattern is in error in certain digit positions. The idea behind error control coding is that, by introducing extra digits into the transmitted signal to provide carefully structured redundancy, it may be possible to detect the presence of errors in the received pattern.

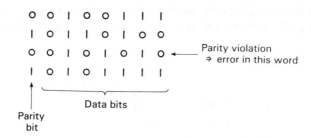

Fig. 7.11   Single error detection using a parity bit.

Indeed, it is seen that it is even possible to so structure the signal that certain errors can be corrected. We shall restrict our attention to binary data since the majority of practical implementations are of this form and also the principle is far easier to appreciate for this case.

Parity checks. Consider the effect of appending an extra bit to the ASCII character codes of Table 7.1 in such a way as to produce an even number of 1s in any character. This extra bit is called a *parity bit* and the 8-bit words representing the characters are said to have *even parity*. If an error occurs during transmission it involves the translation of a 1 to a 0 or a 0 to a 1. In either case, provided no more than one error occurs in any character, the result is a *parity violation*; an erroneous bit results in the associated 8-bit character having *odd parity* and the character is thus known to be in error. The appending of a single parity bit thus provides for single error detection, as illustrated in Fig. 7.11. In order to enable erroneous data to be corrected a return path can be provided, allowing the receiver to send a signal back to the transmitter requesting re-transmission. This can be implemented on a per-character basis or the data can be sent as blocks of characters and the whole block accepted or rejected in accordance with a parity check.

This mode of operation is referred to as ARQ, standing for Automatic Repeat reQuest.

Fig. 7.12 Character-based sum check.

Sum check. An extension of the above idea, widely used in computer communications, considers the sum of the characters viewed as binary numbers. If the number representation adopted admits both positive and negative values a group of characters can be appended to the message to render the total sum zero. This technique, illustrated in Fig. 7.12, is known as a *sum check*; transmission errors will, with very high probability, cause the sum check to fail. There is effectively no limit to the size of the block of characters which may be checked in this way since modular addition can be employed.

Row and column parity checks. The parity checking scheme can also be extended to provide for error correction. One way to achieve this is to consider the transmission of groups of characters and to imagine these organized in a table. Adding a parity bit to each row of the table allows us to check for errors in individual characters but if a parity bit is added for each column then it is also possible to identify the bit position in which an error occurs. With this arrangement, known as row and column parity checking, single errors can be detected and corrected, as illustrated in Fig. 7.13. Note that certain multiple errors can also be detected but not corrected.

This is a simple example of an error detection and correction scheme (EDC). This mode of operation is referred to as FEC standing for Forward Error Control, since no return channel is required.

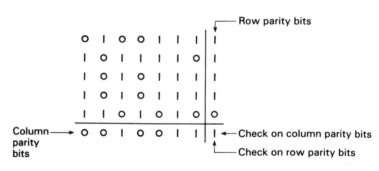

a

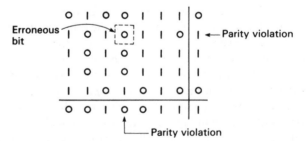

b

Fig. 7.13 Row and column parity checking. (a) No errors; (b) single error detection and identification.

122

Block coding. As a further example of error detection and correction (EDC) coding, appending a number of *check* bits to a block of information bits is considered. This technique is referred to as *block coding* to distinguish it from *convolutional coding* in which the check digits are mixed in with the information bits in a complicated, more or less continuous manner. Specifically, the (7,4) Hamming code is discussed, in which each block contains a total of 7 bits: 4 information bits and 3 check bits. This code provides for both the detection and correction of single errors; the check bits are generated from the data using modulo-2 addition, corresponding to the exclusive-or operation. Consider the data word structure:

[d1   d2   d3   d4   c1   c2   c3]
information   *check*
or data bits   *bits*

One of the earliest papers on this subject is Hamming, R.W., "Error detecting and error correcting codes", *Bell Systems Technical Journal*, **29**, April 1950, 147–160. It is very readable, requiring no advanced mathematical skills.

The check bits are generated from the data as follows:

$$c_1 = d_1 \oplus d_2 \oplus d_4 \qquad (7.12a)$$
$$c_2 = d_1 \oplus d_3 \oplus d_4 \qquad (7.12b)$$
$$c_3 = d_2 \oplus d_3 \oplus d_4 \qquad (7.12c)$$

Here $\oplus$ denotes addition modulo-2 and corresponds to an exclusive-or operation:

$0 \oplus 0 = 0$
$1 \oplus 0 = 1$
$0 \oplus 1 = 1$
$1 \oplus 1 = 0$

and these same equations are used to check the correctness of the data at the receiver. If during transmission any one of $\{c_1,d_1,d_2,d_4\}$ becomes corrupted then at the receiver Equation 7.12a does not hold. Similarly for $\{c_2,d_1,d_3,d_4\}$ and $\{c_3,d_2,d_3,d_4\}$ in relation to Equation 7.12b and c. Consider, for example, that d3 is in error, so that test 7.12a succeeds but 7.12b and 7.12c fail. Since 7.12a is satisfied it is known that $\{c_1,d_1,d_2,d_4\}$ are all correct. Hence the failure of 7.12b is caused by c2 or d3, $\{d_1,d_4\}$ having been validated by 7.12a. But 7.12c fails, so the error cannot be c2 since this is not involved in 7.12c; it is concluded that d3 is in error. d3 is now corrected by complementing it, that is by setting it to 1 if 0 was received and to 0 if 1 was received.

A reordering of the codeword bits of the form

[c1,c2,d1,c3,d2,d3,d4]

is often used since an appropriate reformulation of the tests of Equation 7.12 then allows the position of an error to be directly specified. To achieve this the idea of an error *syndrome* [S] = (s1,s2,s3) which specifies the location of an error is introduced. [S] = (0,0,0) means no error, [S] = (0,0,1) says that c1, the first bit, is in error, etc. The error syndrome is constructed as follows:

The *syndrome* indicates both the presence and the location of an error.

$$[S] = \begin{cases} s_1 = c_3 \oplus d_2 \oplus d_3 \oplus d_4 \\ s_2 = c_2 \oplus d_1 \oplus d_3 \oplus d_4 \\ s_3 = c_1 \oplus d_1 \oplus d_2 \oplus d_4 \end{cases} \qquad (7.13)$$

Here Equation 7.13 is obtained from Equation 7.12a by noting that

$$c_1 = d_1 \oplus d_2 \oplus d_4$$
$$\Rightarrow \quad c_1 \oplus d_1 \oplus d_2 \oplus d_4 = 0 \qquad (7.14)$$

and similarly for 7.12b, c. Here use is made of the result that addition modulo-2 and subtraction modulo-2 are identical operations. Applying Equation 7.13 to the

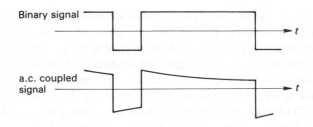

Fig. 7.14 Binary signal with 'droop' due to a.c. coupling.

example where d3 is in error we obtain [S] = (1,1,0), specifying, correctly, that the error is in bit 6.

This general idea can be extended to larger blocks and to the detection of multiple errors per block but this is not considered in this text. Of course, these schemes only work satisfactorily if the number of errors in a block is limited. If errors are likely to occur closely together, in bursts, yet the average error rate is still reasonably low, then EDC can be achieved by distributing check and corresponding data digits throughout the transmitted sequence such that they are widely separated. By this means even if several adjacent bits are corrupted the information can be reconstructed from the uncorrupted check and data digits elsewhere in the data stream. There are many further aspects of error control coding but the above should suffice to illustrate the underlying principle. Broadly speaking we are concerned with deliberately introducing redundancy — extra bits — in such a way that errors that occur during transmission can be detected and/or corrected. This is in marked contrast to source coding where the aim is to remove unwanted redundancy so as to reduce the average number of bits required to represent a message. It is important to realize, however, that the two ideas are *complementary* rather than *contradictory*! A source may contain considerable natural redundancy and yet this may not be in a form which allows for ready and reliable error detection and correction. In these circumstances it may be appropriate to employ source coding to remove the unwanted, *unuseable*, redundancy and to follow this by EDC coding to provide immunity to channel errors. The resultant data rate might be the same, greater, or less than that of the uncoded source signal and yet improved information transfer may result.

Line Coding

Both source coding and error control coding are concerned with matching the source to the channel. With source coding we seek to make the source compatible with a lower data rate channel than might otherwise be required; with error control coding we seek to obtain reliable communication despite a certain degree of unreliability in data transfer over the channel. Line coding is concerned with yet another aspect of matching a source to an available channel.

It is frequently the case that a communication link cannot transmit d.c. components. This is the case in the telephone network, for example, in which a.c. coupling is provided via transformers. In these circumstances, if it is attempted to send a long string of 1s or 0s, the signal 'droops', as shown in Fig. 7.14, leading to discrimination difficulties at the receiver. A line code employs redundancy of some form to eliminate this problem.

Line codes for telecommunications are usually designed such that the power spectrum of the coded digital signal has a null at d.c.

124

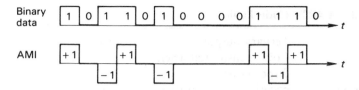

Fig. 7.15 Alternate mark inversion.

Perhaps the simplest example is provided by alternate mark inversion (AMI) in which the binary (2-level) data are converted into a ternary (3-level) signal by representing alternate 1s — or marks — by positive and negative pulses, zeros being left unchanged. This is illustrated in Fig. 7.15; the result is a signal which is balanced around zero. This arrangement avoids the need to transmit a d.c. component but long runs of 0s can still be experienced and the absence of frequent transitions in the data means that the receiver is likely to get out of step with the transmitter. For example, if the receiver clock frequency is 11/10 times that of the transmitter then a run of 10 zeros is interpretted as 11 zero's. To avoid this problem it is required that the line signal should contain frequent transitions from which the receiver clock frequency and phase can be continuously adjusted. A widely used modification of AMI which achieves this is known as HDB_3 — high density bipolar code with substitution after three consecutive zeros. The coding rules are as for AMI unless this would result in more than three consecutive zeros at the output, in which case a '1' is inserted but in such a way as to violate the alternating mark sequence. It can thus be recognized as a violation at the receiver and removed to restore the original data. However, these violations could upset the balance of the line signal so they must themselves alternate and yet be distinguishable from data marks. The result is a rather complex set of coding rules which will not be detailed here. Note, however, that HDB_3 has been adopted by the CCITT as a standard code for interfacing 30 + 2 channel (2 Mbit/s) PCM systems.

An alternative, more efficient form of line coding operates on blocks of data bits rather than on the individual digits. An example is provided by the 4B3T codes in which groups of four binary digits are mapped on to words of three ternary digits. Wherever possible balanced output words, whose digits sum to zero, are employed. For example, the ternary word +1,0, −1 has a zero digit sum and is said to be a *zero disparity* word, while +1, +1, +1 has a digit sum, or disparity, of +3. Nonzero disparity words are grouped in pairs, one positive and one negative, and the positive or negative code word is transmitted as appropriate in order to reduce towards zero the accumulated disparity or running digital sum. An illustrative section of a 4B3T code translation table is shown in Table 7.3. Note that the output codeword 000 is not allowed, although it is zero disparity, since it contains no transitions.

We have discussed ternary line codes since these have found most widespread application in the past. With the advent of optical fibre communications, however, there is an increased need for purely binary line codes. This is because a laser or light emitting diode source can be readily switched on and off to produce pulses of light; a pulse of light can then represent a 1 and no pulse a 0. Since the intensity of the source is being modulated it is not possible, however, to have *negative* light! That is not to say that ternary transmission is not possible — levels equivalent to 0,

Frequent data transitions facilitate the recovery of a 'clock' signal.

Various line codes and their properties are discussed in more detail in Bylanski, P. and Ingram, D.G.W., *Digital Transmission Systems*, Peter Peregrinus, 1976, Chapter 11.

Note that this is just one example of a 4B3T line code — there are many other possibilities.

Table 7.3 A 4B3T Code Table

Binary input	Ternary output Code transmitted when disparity is: negative	positive
0000	+ 0 −	+ 0 −
0001	− + 0	− + 0
0010	0 − +	0 − +
0011	+ − 0	+ − 0
0100	+ + 0	− − 0
0101	0 + +	0 − −
0110	+ 0 +	− 0 −
0111	+ + +	− − −
1000	+ + −	− − +
1001	− + +	+ − −
1010	+ − +	− + −
1011	+ 0 0	− 0 0
1100	0 + 0	0 − 0
1101	0 0 +	0 0 −
1110	0 + −	0 + −
1111	− 0 +	− 0 +

$+1$ and $+2$ could be used — but it is inconvenient and, it transpires, makes less effective use of the power capability of the optical source. A binary line code involves mapping a group of n input bits into $n + m$ output bits. In the interests of efficiency we would like m to be as small as possible to avoid transmitting more bits than necessary. In practice this is invariably achieved by making n odd and $m = 1$. Examples of common binary line codes are 1B2B, 3B4B, 5B6B, 7B8B. The first two are illustrated in Table 7.4. Note that $n,n + 1$ codes are based on n odd since only with $n + 1$ even can zero disparity output words exist.

Table 7.4 Illustrative Binary Line Codes

1B2B Input	Output	3B4B Input	Output + ve	zero	− ve
0	0 1	000	1101		0010
1	1 0	001		1001	
		010		1010	
		011		0011	
		100		1100	
		101		0101	
		110		0110	
		111	1011		0100

Details of other line codes, including extensive code tabulations and graphs of the power density spectra for the coded signals are provided in Cattermole, K.W. and O'Reilly, J.J., *Problems of Randomness in Communication Engineering*, Pentech Press, 1984.

Digital Modulation

A digital signal can be conveyed over a bandpass channel with the aid of modulation using schemes similar to those discussed for analogue messages in Chapter 3. This is another aspect of matching the information source to the available physical channel and may be used in conjunction with the various coding schemes discussed above. The three main forms of digital modulation are only considered briefly.

Amplitude Shift Keying

This is a form of amplitude modulation in which a sinusoidal carrier is switched on for a binary 1 and off for a 0. The transmitted signal has the form

$$x(t) = \left\{ \sum_n a_n \, \text{rect}[(t - nT)/T] \right\} \cos(2\pi F_c t)$$

where F_c is the carrier frequency, $1/T$ is the signalling rate and $a_n = 0$ or 1 for binary signalling. The scheme is readily extended to multi-level signalling by allowing other values for a_n. Fig. 7.16a provides an illustration of the binary case.

Phase Shift Keying

Here the carrier phase is switched in sympathy with the data; Fig. 7.16b provides an illustration. Multi-phase signalling is common and a phasor diagram for a 4-phase case, denoted 4ϕPSK, is shown in Fig. 7.17. Here each of four possible

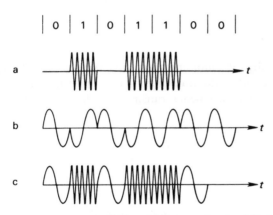

Fig. 7.16 Digital modulation schemes. (a) Amplitude shift keying (ASK); (b) phase shift keying (PSK); (c) frequency shift keying (FSK).

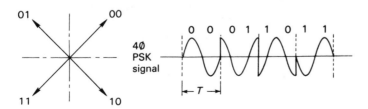

Fig. 7.17 Four-phase PSK (4ϕPSK). (a) Signal vector diagram; (b) illustrative time domain signal. Signalling rate = $1/T$ baud, data rate = $2/T$ bit/s.

signal phases is used to represent a group of two binary data digits. As a result the binary data rate is twice the signalling rate (2 bits per symbol). It is convenient to distinguish the binary data rate, measured in bit/s, from the channel signalling rate, measured in baud: 1 baud corresponds to 1 signalling element per second.

Frequency Shift Keying

An m-ary system conveys $\log_2 m$ bits per symbol. For a binary system $m = 2$, corresponding to $\log_2 2 = 1$ bit per symbol.

Yet another alternative is to switch the instantaneous frequency of the carrier, as shown in Fig. 7.16c. Once again m-ary signalling is possible and in this instance it corresponds to multifrequency operation, which is widely used.

Combined Modulation Schemes

It is perfectly possible to combine the various modulation schemes, a particularly common format being combined multiple-amplitude and multiple-phase shift keying. Fig. 7.18 provides an illustrative *signal space* diagram in which the various points represent possible signal values. Such a diagram is referred to as a *signal constellation*. The primary motive for using these more complex schemes is that they make it possible to achieve a high data rate on a narrowband channel. Broadly speaking, the bandwidth required is related to the symbol rate while the data rate is the symbol rate multiplied by the number of bits per symbol. Hence, if a constellation contains 2^N points, corresponding to N bits per symbol, a data rate of N/T bit/s is achieved while signalling at $1/T$ baud over a channel with a bandwidth of the order of $1/T$ Hz.

Summary

Various aspects of digital communication have been discussed. We began with a description of baseband digital transmission principles and showed how the signal is related to the data sequence being transmitted. The undesirable phenomenon of

Combined amplitude and phase modulation is often referred to as QAM, standing for Quadrature Amplitude Modulation, since the signal can be constructed by adding together two multi-level AM signals, one on a cosine carrier and the other on a sine carrier.

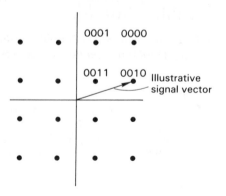

Fig. 7.18   Signal constellation for a simple combined amplitude and phase digital modulation scheme.

intersymbol interference was described and the eye diagram was introduced as a convenient and practically useful means of assessing ISI. Some aspects of signal design for low ISI were discussed briefly. Error probability was introduced as a measure of performance and the process whereby additive Gaussian noise may induce errors was described and quantified. Discrete channel models allowing for finite error probability were discussed.

Several facets of coding for digital transmission were discussed: Source coding involves removing unwanted redundancy from messages; error control coding involves the controlled introduction of redundancy to protect the message data from errors in digital transmission; line coding involves controlled introduction of redundancy to match the digital signal to the physical characteristics of the transmission medium (e.g. zero d.c. content) and facilitate synchronization of the terminal equipment. Digital signals may be conveyed over a bandpass channel by modulating a sinusoidal carrier, and some of the more elementary digital modulation schemes were described.

Problems

7.1 A digital transmission system employs a signal element waveform at the input to the decision circuit given by

$$p(t) = \frac{[1 - \cos(2\pi t/T)]}{2(\pi t/T)^2}$$

Show that this provides zero ISI for a binary signalling rate of $1/T$ bit/s.

7.2 It is required to check the error performance of an 8 Mbit/s binary data transmission system. Approximately how long does it take to make the measurement, with reasonable precision, if the error probability is of the order of 10^{-10}?

7.3 Consider a source which produces messages based on three symbols: A, B, C. The symbols are found to occur, on average, with relative frequencies A = 0.5, B = 0.25, C = 0.25. It is required to encode messages in a binary format; two coding schemes are proposed:

Scheme 1	Scheme 2
A = 01	A = 1
B = 10	B = 01
C = 11	C = 00

Show that, on average, a smaller number of bits per message are required for scheme 2 than for scheme 1.

7.4 A digital transmission system uses the (7,4) Hamming code to provide error detection and correction with the check digits positioned as shown in Equation 7.12. If the binary data pattern 1010111 is received determine whether it is correct or erroneous and, if the latter, the location of the error.

7.5 A combined amplitude and phase modulation scheme has the signal constellation shown in Fig. 7.19. Determine (i) the maximum number of bits which can be represented by each symbol; (ii) the signalling rate required to achieve an information rate of 64 kbit/s; and (iii) the approximate transmission bandwidth required.

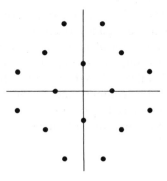

Fig. 7.19   Signal constellation for Problem 7.5.

Systems Case Studies 8

Objectives

☐ To illustrate the technique and principles discussed in previous chapters with reference to specific systems of practical importance.

☐ To describe the system adopted in the UK for FM monophonic and stereophonic sound broadcasting.

☐ To describe the monochrome and colour television systems adopted in the UK.

☐ To show how the design of a television system is strongly influenced by the characteristics of human vision.

☐ To describe briefly some of the multi-media telematic services such as videotex — encompassing teletext and viewdata systems — and facsimile.

☐ To introduce the Compact Disc (CD) digital audio system; interpreting this as a communication system.

In this chapter several illustrative communication systems are examined in some detail. The particular examples included here have been chosen both for their practical importance and because they provide useful illustration of the various techniques and principles discussed earlier.

Broadcast FM Radio

FM radio broadcasting in the United Kingdom is concentrated in the VHF region of the spectrum, based on carrier frequencies in the range 88 MHz to 97.6 MHz. Other carrier frequencies are used elsewhere in the world and to allow for this a receiver usually tunes from 88 MHz to 108 MHz. Radio stations are assigned carrier frequencies at 200 kHz intervals within this band. Signals received beyond the line of sight distance are weak and prone to fading and this, together with the characteristic of FM receivers to suppress weak interfering signals in the presence of a strong intended signal, serves to define the transmitter broadcast area. In allocating channels (operating carrier frequencies) due account is taken of the geographical distribution of transmitting stations together with the transmitter broadcast area. A carrier stability of not worse than 0.002%, approximately 2 kHz, is required by international agreement.

Monophonic FM radio provides for high quality audio signal reception with a message bandwidth extending to 15 kHz. A peak frequency deviation of 75 kHz is employed so that the required receiver bandwidth, by Carson's rule, is given approximately as follows:

$$B_T = 2(f_d + W) = 2(75 + 15) \text{ kHz}$$
$$= 180 \text{ kHz.} \tag{8.1}$$

Carson's rule for FM system bandwidth is discussed in Chapter 3.

Hence with a 200 kHz channel spacing there is a guard band of 20 kHz between adjacent channels to allow for imperfect filtering.

In the receiver, a relatively high IF of 10.7 MHz is commonly employed to ensure good image channel rejection. Ceramic or other high selectivity filters may

131

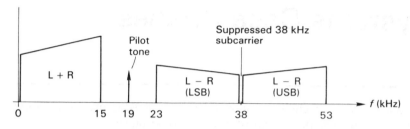

Fig. 8.1   Composite stereo baseband signal spectrum.

be incorporated in the IF stage to provide good adjacent channel rejection. Alternatively, a double conversion superhet receiver may be employed with a first IF at 10.7 MHz to facilitate image channel rejection and a second IF at 465 kHz to facilitate adjacent channel rejection.

FM Stereophonic Broadcasting

FDM is discussed in Chapter 3.

Stereophonic reproduction of sound makes use of two audio channels, designated left (L) and right (R). For stereophonic broadcasting the two audio signals need to be combined in such a way that following transmission to the receiver they can be separated for presentation to two disjoint audio amplifier-and-loudspeaker systems. Also, when the FM stereo system was being devised there was — and indeed still is — a requirement that the stereo signal should be compatible with any existing monophonic receivers. This was accomplished by providing at the transmitter a composite baseband signal of the form shown in Fig. 8.1. This is essentially an FDM signal comprising (L + R) and (L − R) signals together with a 19 kHz pilot tone. The combined left plus right signals, corresponding to a monophonic signal, are concentrated in the frequency range below 15 kHz. This is termed the *sum* signal, L + R. A *difference* signal, L − R, is produced and, with the aid of double sideband suppressed carrier modulation, is translated to the frequency range 23 kHz–53 kHz. It is centred on the (suppressed) carrier frequency of 38 kHz. The latter is referred to as a *subcarrier* of 38 kHz to distinguish it from the FM carrier frequency used for transmission. A low-level 19 kHz signal in phase synchronism with the suppressed 38 kHz subcarrier is introduced. This is termed a *pilot tone* and is provided to facilitate detection of the DSB–SC L − R signal. The *composite stereo signal* is thus an FDM signal comprising the (L + R) and (L − R) signals together with a 19 kHz pilot tone. It occupies the frequency range up to 53 kHz and may be treated as a *composite baseband message signal*. It may thus be frequency modulated in the normal way on to a VHF carrier. Once again, a peak deviation of 75 kHz is employed. An illustrative FM stereo encoder is shown in Fig. 8.2.

See Chapter 4 for a discussion of receiver principles.

At the receiver the composite signal is first recovered by using conventional FM receiver principles. The receiver bandwidth required is slightly greater for stereo than for mono since the composite baseband message has a bandwidth of 53 kHz compared with 15 kHz for a mono audio signal. Applying Carson's rule:

$$B_{\text{T stereo}} = 2(f_d + W)$$
$$= 2(75 + 53) \text{ kHz}$$
$$= 256 \text{ kHz}$$

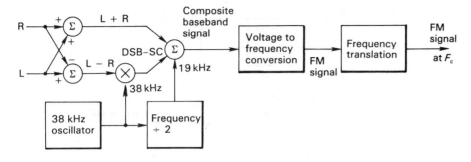

Fig. 8.2   FM stereo encoder and transmitter.

Note that this is slightly greater than the normal channel spacing for FM radio, so care is needed when allocating channel frequencies to transmitters in the same geographical locality. Following FM detection of the composite signal, it must then be decoded and the separate L and R signals recovered. There are various methods whereby this can be achieved; one is illustrated in Fig. 8.3. In this method the L + R signal is recovered by way of a 15 kHz lowpass filter, as in a mono receiver. A bandpass filter centred on 38 kHz and passing signal frequencies in the range 23 kHz to 53 kHz separates out the DSB–SC signal carrying L – R information. The L – R signal is then recovered by synchronous detection and lowpass filtering, as discussed in Chapter 3. This makes use of a phase synchronous 38 kHz local oscillator, phase synchronism being achieved by phase locking to the pilot tone a 19 kHz signal derived from the local oscillator. The separate left and right channel signals are then obtained by adding and subtracting the (L + R) and (L – R) signals:

A phase locked loop (PLL) provides a convenient means of producing from the 19 kHz pilot tone the 38 kHz sub-carrier replica required to demodulate the L – R signal.

$$(L + R) + (L - R) = 2L$$
$$(L + R) - (L - R) = 2R \qquad (8.2)$$

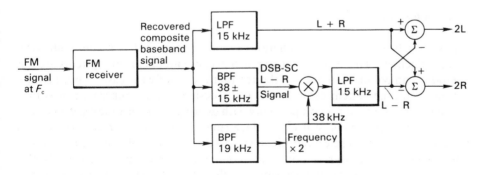

Fig. 8.3   FM stereo receiver and decoder.

The effectiveness of a stereo system depends on the separation achieved between the two channels. Slight imbalances in signal levels tend to result in some crosstalk between left and right channels. Separation can be improved by *matrixing* the two signals, as shown in Fig. 8.4. If the decoded signals are expressed as

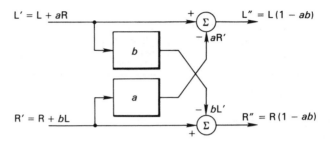

Fig. 8.4   Matrix to improve stereo separation.

$$L' = L + aR \qquad (8.3\text{a})$$
$$R' = R + bL \qquad (8.3\text{b})$$

where $a,b \ll 1$ represent crosstalk, a multiplied by Equation 8.3b is subtracted from Equation 8.3a and b Equation 8.3a from Equation 8.3b to obtain

$$
\begin{aligned}
L'' &= L + aR - a(R + bL) \\
&= L - abL = L(1 - ab) \simeq L
\end{aligned}
\qquad (8.4\text{a})
$$
$$
\begin{aligned}
R'' &= R + bL - b(L + aR) \\
&= R - abR = R(1 - ab) \simeq R
\end{aligned}
\qquad (8.4\text{b})
$$

The matrix may also allow for any differences in level between the decoded left and right channel signals. It should be noted, however, that the crosstalk is generally frequency dependent so perfect separation is not obtained with a frequency independent matrix of the form shown above.

Exercise 8.1 A stereo decoder of the form shown in Fig. 8.3 is found to produce poor separation. On investigation it is found that the 38 kHz local oscillator in the receiver has a phase error of 5°. Suggest a matrixing scheme to correct for this and determine appropriate settings for the adjustable parameters.

Television Systems

This section begins with an examination of monochrome, or black-and-white, television principles. This is done not so much because monochrome systems are less complex than colour systems — although that is certainly the case — but because the monochrome system came first and the design of a colour system was thus influenced by the need for compatibility with existing monochrome receivers. And indeed, this requirement still exists since small monochrome receivers are still produced and sold in large quantities owing to their considerably lower complexity (and hence price) and great portability compared with colour receivers. In this respect the development of television parallels that of FM broadcast radio discussed earlier in which the stereophonic system is required to be compatible with the lower complexity/cost monophonic receivers.

The specific descriptions of television principles presented here relate to systems in use in the United Kingdom, which conform to CCIR recommendations. Slightly different schemes have been adopted in some other parts of the world (notably the United States of America) but these differences are largely of implementation rather than of fundamental principle.

CCIR (Comité Consultatif International de Radiocommunications) — International Consultative Radio Committee.

134

Monochrome Television

A visual scene may be mapped into an electrical signal waveform by way of a raster scanning process, as illustrated in Fig. 8.5. Generally this is achieved by imaging the scene on to a photosensitive surface which is continuously scanned by an electron beam. A current is detected with intensity dependent on the brightness of the image at each point in the scanned field, and this current forms the electrical signal. Note that the scanning process provides for the mapping of two-dimensional information (a picture) on to a one-dimensional coordinate (time). At the receiver an inverse process is employed — scanning is used to convert the one-dimensional signal into a two-dimensional field. The displayed picture inevitably contains a line structure associated with the scanning process. Also, since it is required to transmit moving pictures the scanning process must be repeated over and over again — one frame at a time — so that the observer is presented with a rapid sequence of pictures. The acceptability of this form of *signal encoding* depends ultimately on the characteristics of the human vision system. The eye–brain system interprets a sufficiently rapid sequence of incrementally changing pictures as continuous motion and provided the lines are sufficiently close together the limited spatial resolution of the eye will render these non-visible. Broadly speaking we conclude that a good television system would have very closely spaced scanning lines and would present picture frames in rapid succession. On the other hand in the interests of spectral efficiency it is required that the encoded signal occupies as small a bandwidth as practicable. These two requirements — accuracy of representation versus spectral efficiency — are such that a compromise is required. Systems are designed such that the signal bandwidth is as small as possible given the need for a subjectively acceptable picture display.

See Pearson, D.E., *Transmission and Display of Pictorial Information*, Pentech Press, 1975, Chapter 2.

The elimination of discernable flicker requires some 40 to 50 frames per second while at desirable viewing distances about 500 raster lines per picture are required. The total number of scan lines employed is 625 but not all of these are displayed since time is required for the scanning spot to return to the top of the screen before beginning a new picture. Also, in practice an interlaced scanning arrangement is employed in which every other line is first displayed and the scanning spot then

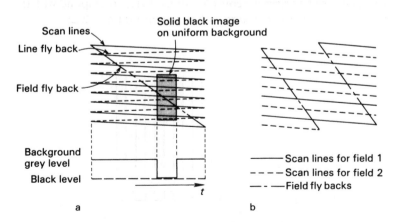

Fig. 8.5 Raster scanning principles. (a) Simple scanning raster and video waveform corresponding to a single line. (b) An interlaced scanning arrangement.

returns to the top of the screen to fill in the gaps on a second pass. There are thus two *fields* per complete picture or *frame*. An interlaced scan is illustrated in Fig. 8.5b. With this arrangement a complete picture is displayed once every 1/25th of a second (25 Hz frame rate) yet, with the 2 : 1 interlace giving a field rate of 50 Hz, picture flicker is acceptably slight.

From the above it can be deduced that every line must be scanned in (1/25)/625 seconds, giving a line rate of 15.625 kHz. Time must be allowed for the spot to fly back from the end of one line to the beginning of another so the *active line time* is only about 52 microseconds. The bandwidth required for a television signal can now be estimated by assuming that the horizontal resolution is required to be comparable to the vertical resolution as set by the number of lines per picture. Allowing for the usual 4 : 3 width to height aspect ratio of a television display and noting that some 50 lines per picture are allocated to frame flyback the process is as follows: The picture is viewed as made up of a rectangular array of picture elements or *pels*. The vertical separation between pels is set by the number of active lines (625 − 50) = 575 and for equal vertical and horizontal resolution we need a similar horizontal spacing. Thus approximately 575 × 4/3 pels per line are required and these must be displayed in the active line time of 52 microseconds. The effective time separation between pels is

$$t_p = 52 \times 10^{-6}/(575 \times 4/3) = 68 \text{ ns}$$

corresponding to a rate of $1/(68 \times 10^{-9}) = 14.7 \times 10^6$ pels per second. The question now arises: what bandwidth is required to be able to transmit pels at this rate? By considering sequences of alternating black and white pels along a line, as shown in Fig. 8.6, it can be concluded that the highest fundamental frequency encountered is given by

$$f_{max} = 1/(2\pi t_p) = 7.3 \text{ MHz}$$

Now the average picture brightness corresponds to a d.c. signal so there is a temptation to conclude that faithful reproduction of a television signal requires a bandwidth extending from 0 to 7 MHz. In fact this is rather pessimistic. An alternative estimate can be based on risetime considerations. Let t_p correspond with the 10% to 90% risetime of a first-order system with time-constant $t_p = 2.2\tau$.

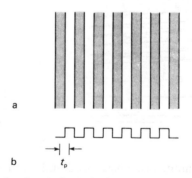

Fig. 8.6   Determining horizontal resolution. (a) Vertical black and white bars with separation equal to 1 pel; (b) video waveform for a single line.

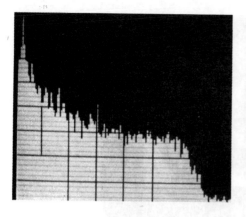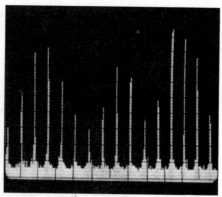

Fig. 8.7   Spectrum of a television signal. (a) Wide scan showing bandwidth requirement; (b) detail showing structure at line rate of 15.625 kHz.

But the 3 dB bandwidth of such a system is given by

$$B = 1/(2\pi\tau)$$

Hence

$$B = 2.2/(2\pi t_p) = 5.1 \text{ MHz}.$$

This might be expected to be an underestimate since filters with a rather sharper cut-off are generally involved. In practice a bandwidth of 5.5 MHz is employed and found acceptable. It may also be noted in passing that the scanned origin of the signal results in a far from uniform distribution in frequency space. Power is concentrated in the vicinity of multiples of the 15.625 kHz line rate as shown in Fig. 8.7.

Timing information is combined with the television signal to enable the receiver to obtain proper line and field synchronization. The resultant composite video signal is shown in Fig. 8.8. The line synchronizing pulses are of 4.7 microseconds duration and occur once every 64 microseconds during the active picture time. A more complicated synchronization pattern is employed during the field flyback interval to allow for interlacing and provide for relatively simple synchronization circuitry within the receiver.

Colour Television

Colour television relies on the principle of additive colour mixing. The wide range of colours required for faithful reproduction of a colour scene can be obtained by adding together appropriate contributions of the *primary* colours: red (R), blue (B) and green (G). The same principles of scanning and interlacing are employed as for monochrome systems but now three separate photosensitive surfaces are required in the television camera; one for each primary colour. A colour separation optical assembly is employed to direct the R,G,B components in a scene to the appropriate camera tube. The images on these tubes are scanned synchronously so that at each time instant R,G,B signals are obtained indicating the amount of red, green and blue required at the receiver to reproduce the colour and luminosity at a corres-

A detailed specification for the 625-line colour television system I is provided in the *IBA Technical Review No. 2*, September 1972.

137

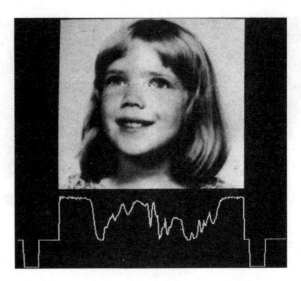

Fig. 8.8   Monochrome video signal incorporating composite synchronization information.

See Pearson, D.E., *Transmission and Display of Pictorial Information*, Pentech Press, 1975, Chapter 8.

ponding point in the image. There are now *three* related video signals to be transmitted and to do this using separate channels for each signal, each having the same bandwidth as for monochrome television, would require a total bandwidth of 3×5.5 MHz = 16.5 MHz. Fortunately it is possible to combine the signals, effecting considerable bandwidth economy and at the same time achieving compatibility with monochrome receivers. The first step is to produce a luminance signal Y corresponding to a weighted sum of the R,G,B components. The weighting employed takes account of the relative luminous efficiencies of the different colour phosphors used in television display tubes. For the purposes of illustration, however, assume a simple summation:

$$Y = R + G + B$$

This luminance signal is the required video signal for monochrome display. The chrominance (i.e. colour) information is then represented by colour difference signals: $R - Y$ and $B - Y$. There are still three signals to transmit but one of them now corresponds to the monochrome signal required for compatibility. Next, if the bandwidth required for *colour* as opposed to *luminance* information is considered, it is found that a considerable reduction is possible.

A -3 dB bandwidth of only 1.3 MHz is employed with chrominance signal level gradually reducing to -20 dB at 4 MHz.

It is now required to combine the three signals:

$$
\begin{aligned}
Y &= \text{luminance, 5.5 MHz bandwidth} \\
\left.\begin{array}{l} R - Y \\ \\ B - Y \end{array}\right\} &= \left\{\begin{array}{l} \text{colour difference signals,} \\ \text{1.3 MHz bandwidth,} \\ \text{gradual roll-off} \end{array}\right.
\end{aligned}
$$

and form a composite video signal such that luminance and colour difference signals can be recovered for colour display, yet the signal is directly compatible with monochrome receivers. To achieve this we first double sideband suppressed carrier modulate the R − Y component on to a cosine colour subcarrier of frequency ∼4.43 MHz and simultaneously modulate the B − Y component on to a corresponding sine subcarrier. These two components are added together to form a quadrature amplitude modulated (QAM) signal. The combined signal has the same spectral occupancy as each component and yet the individual modulating signals (R − Y), (B − Y) are readily recovered using synchronous demodulation with appropriately phased carriers.

Note: QAM and SSB have equal spectral efficiency. A message of bandwidth W can be transmitted in a passband of $B = W$ using SSB while two messages each of bandwidth $W/2$ can be transmitted in a passband of $B = 2\,(W/2) = W$ using QAM.

Exercise 8.2

The two colour difference signals (R − Y), (B − Y) may be viewed as separate messages $m_1(t)$ and $m_2(t)$ modulated on to quadrature carriers, $\cos(2\pi Ft)$, $\sin(2\pi Ft)$. Show that the messages may be recovered at the receiver provided accurate quadrature carrier replicas are available for synchronous demodulation.

Referring to the previous discussion on monochrome television, recall that the spectrum of the luminance signal is concentrated at multiples of the line rate of 15.625 kHz with very little signal power at intermediate frequencies. The chrominance subcarrier frequency has been selected such that the modulated chrominance signal spectral components fall into these gaps in the luminance spectrum. The baseband luminance and chrominance signals may be added to form the required composite colour video signal as illustrated in Fig. 8.9.

In Chapter 3 it was seen that synchronous demodulation of a DSB–SC signal requires a carrier replica to be available at the receiver. This carrier replica must have not only the correct *frequency* but also the correct *phase* relationship to the (suppressed) carrier associated with the signal to be detected. With QAM the demands on phase accuracy are considerable since an error results in a fraction of the R − Y information being demodulated as B − Y and vice-versa. This would give rise to colour (hue) errors in the displayed picture. To ease these difficulties an accurate subcarrier reference is transmitted as a short burst at the beginning of each line. The oscillator in the receiver uses the burst as a reference to correct for any phase errors. In addition the phase of the burst is made to alternate on successive lines. The details of this phase alternation are not dealt with in this text but note simply that it is such that any residual phase errors in the channel tend to give rise to errors of colour saturation rather than of hue. The latter have been found to be subjectively more objectionable than the former. We note also that this *phase* variation on *alternate lines* gives rise to the usual name for this particular colour television system: the PAL system. An illustrative line of a composite colour video signal is shown in Fig. 8.10. The varying amplitude and phase chrominance information is seen to be superposed on the luminance signal. A monochrome

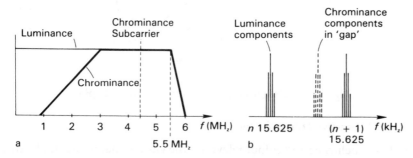

Fig. 8.9 Spectrum of composite colour video signal. (a) General view; (b) detail showing interleaving of luminance and chrominance components.

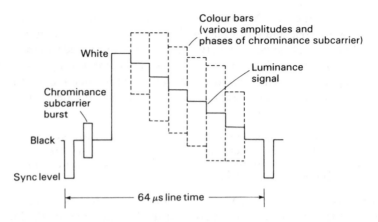

Fig. 8.10   Composite video signal for a colour bar display.

receiver essentially responds to just the luminance component although the chrominance subcarrier components do sometimes give rise to visible patterning, depending on picture content. A colour receiver, on the other hand, can separate luminance and chrominance signals using a delayline filter and recover the individual colour difference signals R − Y, B − Y using synchronous detection. Having recovered Y, R − Y and B − Y, three separate signals R′,G′,B′ can be produced corresponding approximately to the original R,G,B signals but modified by the different bandlimiting applied to the luminance and colour difference signals. These R′,G′,B′ signals are then used to control the currents in three separate electron beams within a television display tube. Each is directed to fall on phosphors of the appropriate colour on the face of the tube as shown schematically in Fig. 8.11.

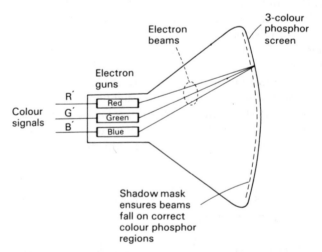

Fig. 8.11   Principle of a colour television display.

Broadcast Television Transmission

For both monochrome and colour television the video signal is essentially lowpass occupying the band $|f| < 5.5$ MHz. In view of this large message bandwidth a spectrally efficient modulation scheme is required; double sideband amplitude modula-

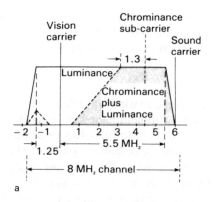
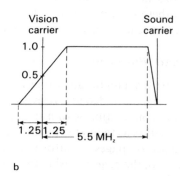

Fig. 8.12 Spectral occupancy of a broadcast television based on
asymmetric/vestigial sideband modulation. (a) Transmitted ASB television
signal plus sound in 8 MHz channel; (b) receiver response to provide ideal VSB
signal.

tion would require a transmission bandwidth of at least 11 MHz per television
channel. Note, however, that the signal contains a d.c. component corresponding
to the average brightness of the picture. This suggests that a single sideband
modulation (SSB) scheme cannot be employed since with SSB d.c. and very low
frequency components in the message are removed by the sideband filter. The
solution lies in the use of an *asymmetrical sideband* (ASB) amplitude modulation
scheme in which most but not all of the lower sideband is removed at the transmitter, as shown in Fig. 8.12a. At the receiver skew symmetric filtering about the
carrier frequency is employed to produce a *vestigial sideband* (VSB) signal as
shown in Fig. 8.12b. Note that the carrier and any d.c. components in the video
message signal are accommodated at half-amplitude. If such a signal is synchronously detected the message waveform is recovered undistorted — the low
frequency attenuation of the upper sideband is compensated by the attenuated
residual (vestige) lower sideband. In practice even if envelope detection is
employed the distortion is not too severe since a strong carrier signal is present. The
associated sound signal is conveyed using frequency modulation of a separate
sound carrier located 6 MHz above the vision carrier. The combined modulated
sound and vision signal thus occupies a bandwidth of 8 MHz per channel.

Videotex Systems

Videotex systems employ alphanumeric characters and graphics on a television
type display to provide for data retrieval. The display is based on a rectangular
array of character cells arranged to fill the screen; data stored within the videotex
terminal governs the character or graphic symbol displayed at any given location
on the screen and also the use of different colours within the display. There are
essentially two general classes: teletext and viewdata. The former is based on
broadcast television while the latter involves the use of two-way communication
over the switched telephone network. In the UK two hardware compatible teletext
systems known as CEEFAX and ORACLE are operated by the television broadcasting

These are examples of *telematic
services* based on a judicious
blend of computer and
communications techniques.

141

stations and a viewdata service known as PRESTEL is provided by British Telecom. Only these systems are considered here, although it should be noted that various different schemes are in use or under investigation in other countries.

Teletext (Broadcast Videotex)

Teletext is a data broadcasting system in which information is carried by bursts of digital signals added to a television video signal. Alphanumerics and graphics are used in conjunction with a television receiver to provide for selective display of information as requested by the user. It is a *one-way* information system in which a sequence of pages of information are sent repetitivily. The user indicates the address of the page he wishes to view by entering the code on a small hand-held keyboard and must then wait until this page appears in the transmitted sequence so that his receiver can capture and display it. A page comprises 24 rows of 40 characters, including a special top row called the page header. This contains address and control data to identify the page and to control its display; it is also used to display general information such as the time, date, page identity, and so on. The rate at which pages can be transmitted depends on how many of the 26 potentially available spare television lines are used for teletext data. When first introduced the system used two lines per field (four per picture) and could then transmit four full pages per second. The system constitutes a form of electronic magazine giving continuously updated news, stockmarket reports, etc. and can also provide subtitles for overlaying on to the normal television picture to help the hard of hearing. We will not consider further the various possible applications of the system; nor will we detail the display format employed. Rather we concentrate on the signal and coding aspects to provide practical illustration of the digital communication principles outlined in Chapter 7.

The data are carried as a non-return to zero (NRZ) baseband binary signal superposed on otherwise unused lines, outside the active picture time, during the field flyback intervals. The relationship between the bursts of data and the normal television signal is shown in Fig. 8.13. The data signal is bandlimited so that the time domain data pulse and its spectrum are approximately as shown in Fig. 8.14. Note that the spectrum has skew symmetry about the Nyquist frequency (3.47 MHz) of

Skew symmetry about the Nyquist frequency provides for zero ISI.

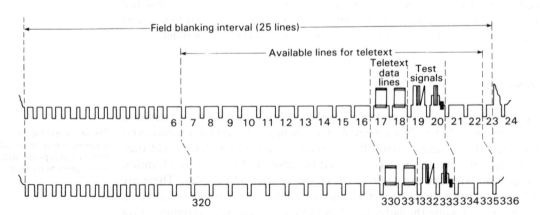

Fig. 8.13   Teletext data lines in field blanking internal.

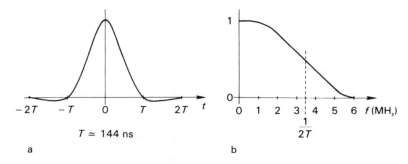

T ≈ 144 ns

$\dfrac{1}{2T}$

a

b

Fig. 8.14 Teletext data pulse and spectrum. (a) Data plus shape; (b) spectrum of data pulse.

one half the signalling rate (6.94 MHz). This is to minimize intersymbol interference as discussed in Chapter 7.

Each data line contains 360 bits of information organized as 45 8-bit bytes. Each line begins with a synchronization pattern: there are 16 bits (2 bytes) of alternating ones and zeros, 101010 . . ., to ease bit-clock synchronization and these are followed by a single byte framing code, 11100100, which provides for byte synchronization. This allows the receiver to determine whether a particular bit is, for example, the last bit of the nth byte or the first bit of the $(n + 1)$th byte. These first three bytes have even parity; the total number of ones is even. The remaining 42 bytes in a line use odd parity and carry address, control and information character data. The use of odd parity ensures that the maximum interval between data transitions is 14 bits. This is essentially a simple form of line code which eases the recovery of bit-clock from data, as discussed in Chapter 7.

It is especially important that address and page control data be received correctly since the user selects information by address and the page control data determines how the information is presented on the television screen. Consequently error detection and correction coding is employed for these data items. This is based on a Hamming code in which each byte contains four message bits and four protection bits. This allows for the correction of both single-bit and multiple-bit errors. If multiple-bit errors are detected then the message is rejected and the receiver waits for its next occurrence in the data sequence.

Perhaps the greatest limitation of teletext is that, being a non-interactive, broadcast system, the average time required to access a page increases with the number of pages included in the transmission sequence. There is thus a definite practical limit to the volume of data to which the user can be given access if the waiting time following page selection is not to be too great. This problem does not arise with a viewdata system.

Use of odd parity with one parity bit per byte facilitates bit synchronization.

Viewdata (Interactive Videotex)

Viewdata is a *two-way* information system based on a combination of television display, local storage and processing, together with access to a remote data store via the public telephone network. Since a two-way communication channel is available each user is able to select specific information and only this specifically requested information is transmitted to the user. As a result the amount of

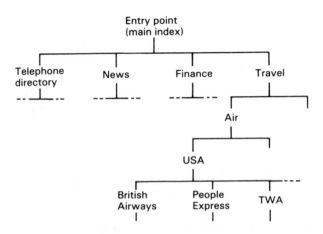

Fig. 8.15 Segment of tree-structured database.

information to which the user can be given almost immediate access is practically unlimited. The information pages stored in the remote database are accessed via a tree search procedure which allows the user to penetrate to increasing levels of detail, as illustrated schematically in Fig. 8.15.

In the UK similar data formats and coding schemes have been adopted for view-data and teletext in the interests of maximum hardware compatibility. The differences lie in the way in which the user selects the information to be displayed and in how the information is conveyed to the user's premises. The public switched telephone network provides an essentially *analogue* communication channel to the user, even if digital transmission is employed within the network itself. Consequently some form of digital modulation is required. This is provide by a data modem (modulator and demodulator) incorporated in the user terminal. Data transmission is effected at a rate of 1.2 kbit/s using frequency shift keying.

Since data transfer is user specific, viewdata can in principle provide an electronic mail facility; messages directed to a particular user or group of users can be stored until that user makes a call into the system, identifies himself or herself and requests messages. This emphasizes once again the major difference between teletext and viewdata. Teletext, being a broadcast system, is well suited to the transmission of a limited volume of information to which a large number of users require access. Viewdata on the other hand is discriminatory: it provides for user interaction so that data can be selectively directed to specific locations.

In due course the introduction of an Integrated Services Digital Network (ISDN) facility will provide direct digital access (DDA) obviating the need for a modem in the user's terminal.

Facsimile

Facsimile provides a means of transmitting information directly from a printed, drawn or handwritten page and is seen as an important facet of the evolving 'electronic office'. It provides a service whereby the need to physically transport letters, documents, and so on, is avoided by electronically transmitting their contents and regenerating the documents at the receiver. It makes use of a scanning system to effect the requisite two-dimension to one-dimension mapping at the transmitter and similarly for the inverse operation at the receiver. In this respect it is broadly similar to television but only static images are involved so there is no need for frequent re-scanning to accommodate motion and avoid flicker. A document is scanned just once, transmitted and reproduced at the receiver. There is thus no fundamental restriction on the length of time taken to scan a page and very high spatial resolution can be achieved over a limited bandwidth channel.

Most facsimile communication makes use of the public switched telephone network and there are a variety of equipments available. The simplest of these (Group 1 machines) provide a resolution of 3.85 lines/mm and use analogue transmission based on double sideband amplitude modulation of a 1.3 kHz or 1.9 kHz carrier. They require some 6 minutes to transmit an A4 page. Some improvement is provided by Group 2 machines which make more effective use of the available telephone channel bandwidth by using vestigial sideband modulation of a 2.1 kHz carrier. This reduces the transmission time to about 3 minutes. Group 3 machines on the other hand are digital. Only black and white documents are transmitted and various coding schemes are adopted to reduce the number of bits required to describe the document. A resolution of 7.7 lines/mm is provided and a transmission rate of 4.8 kbit/s is employed. Coding is performed on a one-line-at-a-time basis and the transmission time required for a line depends on the information density in that line. As a result the transmission time depends on the type of document involved but typically an A4 page can be transmitted in less than one minute.

A coding scheme which can be used to reduce the number of bits required to describe a document is considered as follows: The document is scanned and a sequence of black and white picture elements is encountered; these are represented by 1s and 0s respectively. These occur in the form of runs of 0s separated by runs of 1s. Rather than transmit the 1s and 0s as such, the *length* of each run of 1s and 0s occurring along each line can be transmitted. This is referred to as *run-length coding*. It makes use of the fact that there is a high degree of correlation between adjacent pels: a 0 is much more likely to be followed by another 0 than by a 1 since most documents consisting of printed or written material are predominantly white space. The general principle of run-length coding is illustrated schematically in Fig. 8.16. This provides a practical illustration of the general principle of *source coding* outlined in Chapter 7. Essentially a notional mathematical model is constructed for the source which takes account of the structure in the data as determined by the type of documents required to transmit. The coding scheme is in some sense matched to the source model and so reduces the number of bits required to represent messages from the source.

Most organizations, including small businesses, now have Group 3 FAX (facsimile) available as part of their general telecommunications provision.

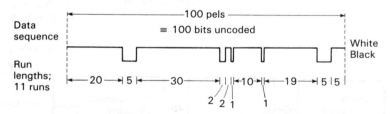

Fig. 8.16   Principle of run-length coding. A fixed 5-bit code per run (maximum run 32 pels) ≡ 55 bits.

One of the consequences of employing source coding is that the messages become less tolerant of transmission errors. For an uncoded picture an isolated bit error would affect just one pel but with a run-length coded picture a single bit error changes the length of a run and so corrupts all subsequent pel information on that line. Error control coding can be used to alleviate these difficulties. It may at first sight seem odd that having reduced the redundancy with the aid of source coding it is now proposed to put back redundancy in the form of channel coding. However, by doing this an *overall* benefit can be achieved. The redundancy required in the error control code to achieve adequately reliable transmission is small compared with the redundancy reduction provided by the source coding operation. The reduction in the number of bits required to be transmitted as a result of the coding operations depends on the type of document and ranges from 17 : 1 for business letters and diagrams to 5 : 1 for very dense text. Other more efficient coding schemes are under investigation but the above should suffice to enable the principles and prospects to be appreciated.

The Compact Optical Disc as a Communication System

Some of the optoelectronics aspects of the CD system are discussed in another book in this series: Watson, J., *Optoelectronics*, Van Nostrand Reinhold, 1988.

The compact optical disc (CD) system for digital recording and reproduction of audio signals can be viewed as a communication system. The studio recording and disc production processes correspond to the transmitter while the CD player represents the receiver. We shall not concern ourselves here with the disc production process nor with the way in which the information stored on the disc in binary form is sensed using a laser to provide the 'received' binary signal. Rather we concentrate on the signal encoding, decoding and processing operations.

The system is shown in outline block diagram form in Fig. 8.17. Two separate

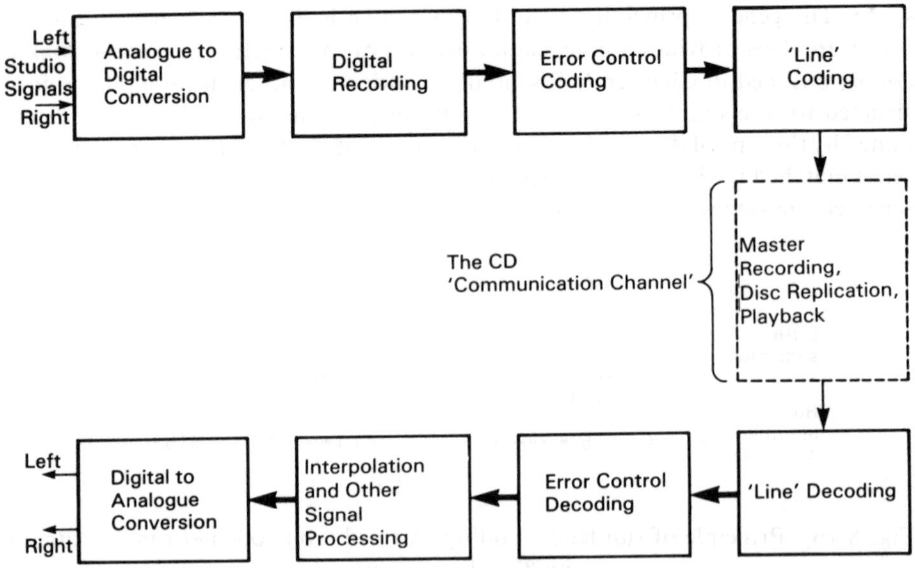

Fig. 8.17 The CD digital audio system, viewed as a communication system.

audio signals are involved, each having bandwidth of approximately 20 kHz, corresponding to the 'Left' and 'Right' channels of a stereo system. To meet the requirements of the sampling theorem and provide some margin to allow for the finite cut-off rate of practical filters each channel is sampled at approximately 44 kHz. Each sample is then converted, using pulse code modulation with uniform quantization, into a 16 bit binary codeword. The binary data rate at the output of the analogue to digital converter (ADC) section is thus approximately $2 \times 16 \times 44$ kbit/s = 1.584 Mbit/s. This represents the basic binary data source. Subsequent digital encoding is required to render the signal compatible with the 'channel' — the disc production and playback process.

The sampling theorem was discussed in Chapter 5.

Pulse code modulation was discussed in Chapter 6.

Certain additional information is also incorporated providing, for example, a track identification facility for the recorded disc, but this need not be considered here.

Error Control

To allow for the possible introduction of errors in the 'channel' some form of error control coding is required. The CD system uses a non-contacting optical readout technique and since the coded disc surface scanned by the laser beam is protected by a transparent plastic layer the signal surface itself is protected from accidental damage such as scratches. Minor defects in the signal surface arise during manufacture and may give rise to random errors covering several bits. The error control code is required to protect against these and also large bursts of errors which may result from fingerprints or scratches above a tolerable level. The FEC code must thus accommodate both isolated errors and error bursts.

The desired error control performance is achieved by using a Cross-Interleaved Reed-Solomon Code (CIRC). This involves the concatenation of two Reed-Soloman (RS) codes arranged together with a set of delays which delay differently the various symbols in a codeword. This ensures that information symbols which are adjacent in the audio signals are widely separated when coded onto the surface of the disc. Consequently errors which occur during reproduction and playback affecting adjacent coded signals *on the disc* become widely separated by the decoding process and can thus be corrected as if they were isolated random errors.

At the encoding stage the first RS code introduces redundancy, translating blocks of 24 information symbols (8-bit Bytes) into 28-Byte codewords. These are then passed via the network of delays to the second RS coder which introduces further redundancy performing a 28-Byte to 32-Byte coding transformation. Following error control coding the signal is passed to the 'line' coder. This introduces redundant bits in order to control the run lengths in the signal recorded on the disc. This facilitates synchronization and renders the signal tolerant to variations in the frequency response of the recording 'channel'. The particular code used maps 8-bit symbols into 14-bit symbols and adds 3 *merging bits* between coded words, providing very tight control of run lengths. This operation is referred to as Eight to Fourteen Modulation coding or EFM coding. The term modulation is used here by analogy with modulation of a sinewave carrier — it renders the message signal compatible with the spectral characteristics of the CD channel.

At the receiver — the CD player — these various coding operations must be 'undone'. First comes the EFM decoding followed by the decoding of the *inner* RS code. At this stage errors may be detected and if there are not too many they are corrected. More emphasis is placed though on detection at this stage. The decoder is so arranged as to forego some of the *error correction* power of the 28–32 RS code in favour of enhanced *error detection* capabilities. Codewords containing more than a very small number of errors are passed forward through the differential delay network but are 'flagged' as erroneous. The constituent symbols are separated by the differential delays so that they appear at the *outer* RS decoder in different 28-symbol blocks. Since these symbols have been detected as suspect by the first decoder they are treated as 'erasures' by the second decoder. The significance of this is that with a given amount of redundancy it is possible to correct twice as many *erasures* as *errors*. Here also, though, some of the decoding power is reserved for just severe error detection — giving up some of the correction capability. Once more, suspect words are flagged to be treated as erasures.

At this stage, in the absence of errors, the audio signal has been recovered, albeit still in digital form. If there are any flagged erasures these are corrected if possible by interpolating between adjacent correct samples. The assumption here is that adjacent samples of a practical audio signal are most likely to be quite similar in amplitude so that interpolation provides a good approximation to the erased value.

The final step is simply the conversion of the digital signal samples into analogue form. Some further signal processing to suppress quantization noise and simplify the post-conversion filtering requirements is sometimes included but we shall not discuss this here. With 16-bit digital to analogue conversion — a 16-bit pulse code modulation system — a signal to quantization noise ratio of approximately 90 dB can be anticipated, assuming a 10 dB peak to mean power ratio for the message signal.

Summary

In this chapter we have described a number of practical communication systems. These are important in their own right but serve also to illustrate further the principles introduced previously. Broadcast FM radio provides an illustration of frequency modulation while stereophonic FM broadcasting involves also double sideband suppressed carrier (DSBSC) modulation and frequency division multiplexing (FDM) principles. Television has been seen to involve a scanning process, a 2-D to 1-D mapping operation, to translate the *message* (the scene or picture) into an *electrical signal* (the TV waveform). Colour television has been seen to make use of DSB–SC and quadrature amplitude modulation (QAM) techniques and to exploit the line structure of the TV signal spectrum to allow colour and luminance information to be combined in a limited bandwidth of approximately 6 MHz. TV transmission makes use of vestigial sideband modulation in the interests of spectral economy.

The combined use of television, computer database, and digital transmission techniques to realize both broadcast and interactive videotex system has been discussed. These provide examples of 'Telematic' systems — i.e. information systems involving a judicious merging of computer and communication techniques. Another telematic service, facsimile, was also considered. Modern, digital facsimile systems make use of run-length coding to remove redundancy from the message. This provides a simple, practical illustration of source coding. Finally we have examined briefly the compact laser disc (CD) system for digital audio, viewing this as a communication system embodying many of the principles and techniques introduced in earlier chapters.

Appendix A

Decibels

The decibel is a logarithmic measure used for comparing two power levels, say P1 and P2. The power ratio P_2/P_1 may be expressed in decibels (dB) as

$$x = 10 \log_{10}(P_2/P_1)\, dB \tag{A.1}$$

We can then say that the power level P_2 is x dB relative to P_1. Note that since $\log(1) = 0$, $x < 0$ implies $P_2 < P_1$ while $x > 0$ implies $P_2 > P_1$.

Equation A.1 may be rewritten in terms of voltages as follows: If P_1 corresponds to a voltage V_1 across a resistor R_1, and P_2 to a voltage V_2 across a resistor R_2, with $P_1 = V_1^2/R_1$, $P_2 = V_2^2/R_2$, then

$$
\begin{aligned}
x &= 10 \log_{10}(P_2/P_1) \\
&= 10 \log_{10}\left(\frac{V_2^2}{R_2} \middle| \frac{V_1^2}{R_1}\right) \\
&= 10 \left\{ \log_{10}\left(\left(\frac{V_2}{V_1}\right)^2\right) + \log_{10}(R_1/R_2) \right\}
\end{aligned}
\tag{A.2}
$$

Assuming $R_1 = R_2$, such that $(R_1/R_2) = 1$, and noting that $\log(1) = 0$, we have

$$
\begin{aligned}
x &= 10 \log_{10}\left[\left(\frac{V_2}{V_1}\right)^2\right] \\
&= 20 \log_{10}(V_2/V_1)
\end{aligned}
\tag{A.3}
$$

The decibel is defined in terms of power ratios but it is not uncommon for voltage ratios to be expressed in terms of decibels, following Equation A.3. This is strictly valid only if $R_1 = R_2$ although on occasions it may be convenient not to adhere rigidly to this condition.

The logarithmic character of the decibel measure for transfer ratio has an important consequence when considering cascade connections of networks. Consider, for example, two networks in cascade, as shown in Fig. A.1, such that

$$
\begin{aligned}
V_1 &= A_1 V_0 \\
V_2 &= A_2 V_1
\end{aligned}
$$

whence

$$V_2 = A_1 A_2 V_0$$

and

$$\frac{V_2}{V_0} = A_1 A_2 = A \tag{A.4}$$

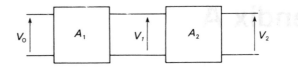

Fig. A.1 Two networks in cascade.

That is, the overall voltage gain is the product of the individual stage gains. Consider now the voltage transfer ratios expressed in decibels:

$$
\begin{aligned}
A_{dB} &= 20 \log_{10}(A_1 A_2) \\
&= 20 \log_{10}(A_1) + 20 \log_{10}(A_2) \\
&= A_{1_{dB}} + A_{2_{dB}}
\end{aligned} \tag{A.5}
$$

That is, we add transfer ratios expressed in decibels for cascade networks.

Finally, we note that in communication systems studies it is often convenient to express absolute power levels in terms to decibels relative to some fixed power reference. For example, the logarithmic measure dBm is used to denote power levels relative to 1 mW and since 100 mW is 20 dB (i. e. 100 times) greater than 1 mW it corresponds to a power level of 20 dBm. For larger power levels dBW, taking 1 W as the reference, is employed. Table A.1 provides some illustrative examples.

Table A.1 Various power levels expressed in watts, dBm and dBW

1 μW	= 10^{-6} W $\equiv$	-30 dBm $\equiv$	-60 dBW
10 μW	= 10^{-5} W $\equiv$	-20 dBm $\equiv$	-50 dBW
100 μW	= 10^{-4} W $\equiv$	-10 dBm $\equiv$	-40 dBW
1 mW	= 10^{-3} W $\equiv$	0 dBm $\equiv$	-30 dBW
1 W	= 10^{0} W $\equiv$	$+30$ dBm $\equiv$	0 dBW
10 W	= 10^{1} W $\equiv$	$+40$ dBm $\equiv$	$+10$ dBW

Appendix B

Some Fourier Transform Results

	Time domain	*Frequency domain*
Definition	$x(t) = \int_{-\infty}^{\infty} X(f)\exp(j2\pi ft)\,\mathrm{d}f$	$X(f) = \int_{-\infty}^{\infty} x(t)\exp(-j2\pi ft)\,\mathrm{d}t$
Reciprocity	$X(t)$	$x(-f)$
Time translation	$x(t-T)$	$X(f)\exp(-j2\pi fT)$
Time scaling	$x(t/T)$	$TX(fT)$
Differentiation	$\dfrac{\mathrm{d}}{\mathrm{d}t}x(t)$	$j2\pi fX(f)$
Convolution	$x(t) * h(t) =$	$X(f)\cdot H(f)$
	$\int_{-\infty}^{\infty} x(u)h(t-u)\,\mathrm{d}u$	
Multiplication	$x(t)\cdot h(t)$	$X(f) * H(f)$
	$\delta(t)$	1
	1	$\delta(t)$
	$\mathrm{rect}(t/T)$	$T\,\mathrm{sinc}(fT)$
	$\mathrm{sinc}(t/T)$	$T\,\mathrm{rect}(fT)$
	$\exp(j2\pi Ft)$	$\delta(f-F)$
	$\cos(2\pi Ft)$	$\tfrac{1}{2}\delta(f\pm F)$
	$\sin(2\pi Ft)$	$-\dfrac{j}{2}\{\delta(f+F) - \delta(f-F)\}$
	$\displaystyle\sum_n \delta(t-nT)$	$\dfrac{1}{T}\displaystyle\sum_n \delta\left(f-\dfrac{n}{T}\right)$
	$\displaystyle\sum_n g(t-nT) =$	$G(f) \times \dfrac{1}{T}\displaystyle\sum_n \delta\left(f-\dfrac{n}{T}\right)$
	$g(t) * \displaystyle\sum_n \delta(t-nT)$	

Answers to Numerical Problems

1.1 (i)   15 km.
   (ii)   75 dB.
   (iii) 1.6×10^{-14} W $\equiv -78$ dBm.
1.2 $+6$ dB $\equiv 4$ times.
1.3 153 links; use a LAN.
1.4 4 Mbit/s.

2.1 (i) $p(t) + p(t - 2T) + p(t - 4T) + p(t - 5T) + p(t - 7T)$.
   (ii) $p(t) + p(t - 4T) + p(t - 8T) + p(t - 10T) + p(t - 14T)$.
2.2 (ii) $x(t) = \sum_n C_n \exp(-j2\pi nt/T)$

   with $C_n = \dfrac{1}{3} \operatorname{sinc}(n/3) = \sin(\pi n/3)/\pi n$

   $$C_0 = \frac{1}{3};\ C_1 = C_{-1} = \frac{1}{2\pi};\ C_2 = C_{-2} = \frac{\sqrt{3}}{2\pi};\ C_3 = C_{-3} = 0;$$

   $$C_4 = C_{-4} = \frac{-1}{8\pi};\ C_5 = C_{-5} = \frac{-\sqrt{3}}{10\pi};\ C_6 = C_{-6} = 0;$$

   $$C_7 = C_{-7} = \frac{1}{14\pi};\ C_8 = C_{-8} = \frac{\sqrt{3}}{16\pi};\ C_9 = C_{-9} = 0.$$

2.3 $Y(f) = X(f) - \dfrac{1}{3}$, i.e. $C_0 = 0$, no d.c. component.

2.4 (i) $x(t) = \displaystyle\sum_{n=-\infty}^{\infty} \delta(t - nT)$.

   (iii) $X(f) = 1/T$.
2.5 (ii) $X(f) = T \operatorname{sinc}^2(fT)$.
2.6 25%.
2.7 (i)   1/8.
   (ii)   1/4.

2.8 (i)   Let $x(t) = \cos(2\pi t)$; then $X(f) = \dfrac{1}{2}\delta(f \pm 1)$,

   i.e. $C_0 = 0$, $C_1 = C_{-1} = \dfrac{1}{2}$, $C_n = 0$ for $|n| > 1$.

   (ii) $S_x(f) = \dfrac{1}{4}\delta(f \pm 1)$.

   (iii) $P = \dfrac{1}{2}$.

   (iv) $P = \dfrac{1}{2}$.

3.1 (i) $\dfrac{1}{3}$.

   (ii) $\dfrac{1}{9}$.

(iii) $\dfrac{a^2}{(2 + a^2)}$.

For speech $\sim 1/101 \cong 1\%$.

3.2 (i)   Two audio tones at $\omega_m + \omega_e$ and $\omega_m - \omega_e$.

(ii)   Single audio tone at $\omega_m + \omega_e$.

Here ω_e is the carrier/local oscillator frequency error. For speech the distortion (splitting) of the message spectrum in (i) is subjectively more annoying then the translation of the message spectrum in (ii).

3.3 96 kHz.

4.1 10.

4.2 (i)   98.6 MHz to 118.6 MHz.

(ii)   109.2 MHz to 129.2 MHz.

(iii) Close to 0 dB (input at 88 MHz has image frequency of 109.2 MHz — this is so close to 108 MHz that there is virtually no image rejection).

5.1 (a) (i) > 216 kHz.

(ii) > 480 kHz.

(iii) > 106 kHz.

(b) (i)   108 kHz.

(ii)   240 kHz.

(iii) 53 kHz.

5.2 -32 dB.

5.3 (i)   240 kHz.

(ii)   240 kHz.

6.1 (i)   6.4 MHz.

(ii)   38.4 Mbit/s.

(iii) 46.77 dB.

(iv)   11 MHz, 88 Mbit/s, 58.77 dB.

7.1 $p(0) = 1$; $p(nT) = 0$ for n integer, $n \neq 0$.

7.2 ~ 35 hours.

7.3 2 bits/symbol for scheme 1, 1.5 bits/symbol for scheme 2.

7.4 Erroneous. Correct data $= 1010101$.

7.5 (i)   4 bits.

(ii)   16 kbaud (16 ksymbols/s).

(iii) 8 kHz.